WELL CEMENT INTEGRITY

WELL CEMENT INTEGRITY

Edited by

Rahman Ashena
Jacqueline Greenwell
William K. (Bill) Ott
Smart Energy Data Analytics (SEDA)-Group, Houston, TX, United States

ELSEVIER

Elsevier
Radarweg 29, PO Box 211, 1000 AE Amsterdam, Netherlands
50 Hampshire Street, 5th Floor, Cambridge, MA 02139, United States

ISBN: 978-0-443-51564-4

For Information on all Elsevier publications visit our website at https://www.elsevier.com/books-and-journals

Publisher: Priscilla Braglia
Acquisitions Editor: Fran Kennedy-Ellis
Editorial Project Manager: Harshita Manchanda
Production Project Manager: Maria Bernard
Cover Designer: Bakyalakshmi S

Typeset by Aptara, New Delhi, India

Dedication

This handbook is dedicated to our beloved families and to all those who supported us and graciously allowed us the time and dedication needed to complete this book.

Contents

4. Sustained casing pressures—Prevention monitoring and remediation

Rahman Ashena, Ali Ghalambor, and Matteo Loizzo

5. Cement bond logs and evaluation

Rahman Ashena and William K. (Bill) Ott

6. Pressure integrity tests

Rahman Ashena

7. Temperature noise and fiber optic logs

Jacqueline Greenwell and Rahman Ashena

8. Corrosion and other logs

Rahman Ashena and Zanko Asadi

9. Fundamentals of cement plugs and evaluation

Rahman Ashena and Razak Yakob

10. Data-driven and machine learning approaches to cement evaluation

Jocin J. Abraham, Mahmood Amani, and Catalin Teodoriu

11. Plug and abandonment

Rahman Ashena, William K. (Bill) Ott, and William Bellenger

12. Cement integrity in carbon capture, utilization, and storage wells

Rahman Ashena, Ibnelwaleed A. Hossein, and Asad Elmgerbi

Contributors

Jocin J. Abraham Texas A&M University at Qatar, Doha, Qatar

Mahmood Amani Texas A&M University at Qatar, Doha, Qatar

Zanko Asadi Independent Consultant

Rahman Ashena Smart Energy Data Analytics (SEDA)-Group, Houston, TX, United States

William Bellenger Smart Energy Data Analytics (SEDA)-Group, Houston, TX, United States

Asad Elmgerbi Montanuniversität Leoben, Leoben, Austria

Ali Ghalambor Oil Center Research, L.L.C., Lafayette, LA, United States

Jacqueline Greenwell Smart Energy Data Analytics (SEDA)-Group, Houston, TX, United States

Ibnelwaleed A. Hossein Qatar University, Doha, Qatar

Matteo Loizzo Independent Consultant, Germany

William K. (Bill) Ott Smart Energy Data Analytics (SEDA)-Group, Houston, TX, United States

Catalin Teodoriu University of Oklahoma, Norman, OK, United States

Razak Yakob Independent Consultant, Kuala Lumpur, Malaysia

Preface

Well integrity has become an increasingly prominent topic—especially over the past decade—as the industry evolves in response to the global push toward sustainability and clean energy. It plays a central role not only in maintaining safe and uninterrupted production but also in enabling the responsible abandonment, repurposing, and long-term management of wells across various sectors. This is particularly vital within the context of the energy transition, where ensuring cement integrity is foundational for the safe deployment of technologies like carbon capture, utilization, and storage (CCUS), as well as both conventional and unconventional geothermal energy.

Well cement integrity stands as the cornerstone of both primary and secondary barriers. It is essential for ensuring well stability throughout the entire life cycle, from drilling and completion to production, abandonment, and repurposing. As new applications place wells under more extreme conditions—such as higher temperatures, pressures, and chemically aggressive environments found in geothermal and CCUS operations—the demand for robust and thoroughly validated cement systems has never been greater.

There can be no compromise when it comes to safety and containment in these evolving energy transition sectors. A holistic approach to well cement integrity is urgently needed—one that considers the dynamic nature of subsurface conditions and integrates all relevant evaluation techniques throughout the life of the well. This book emphasizes that necessity and delivers practical solutions to meet it.

Unfortunately, comprehensive and consolidated knowledge on this topic has not been widely available in existing literature. This book was developed to fill that critical gap. It offers a detailed, transparent, and practical guide to cement integrity evaluation—from laboratory methods to field diagnostics—designed to support well integrity management in both traditional and emerging energy domains.

The book addresses a full spectrum of evaluation methods, including experimental lab testing, sustained casing pressure (SCP) detection, sonic and ultrasonic cement bond logs (CBL), log interpretation with quality control measures, positive pressure testing (PPT), negative pressure and inflow testing (NPT), temperature and noise logs, fiber-optic sensing, cement plugs, and plug and abandonment (P&A) integrity protocols. Special attention is given to cement integrity in high-pressure, high-temperature

(HPHT) wells, and more demanding CCUS environments, where long-term zonal isolation is critical. In addition, a dedicated chapter explores how artificial intelligence and machine learning are driving new capabilities in cement integrity assessment and predictive analytics.

With practical guidance, case studies, and hands-on tools, this book serves as an essential reference for engineers, researchers, and practitioners involved in well integrity, cementing operations, repurposing projects, and energy transition developments. Whether you are working in petroleum, geothermal, CCUS, or related fields, the insights in this book are designed to be applied directly in real-world scenarios.

This book also reflects on recent advancements that have addressed previous limitations in cement integrity evaluation and outlines future directions for improving industry practices. It advocates for continued collaboration between industry and academia, which remains key to driving innovation and addressing the complex challenges of well integrity.

Several practical case studies are included throughout the book to provide clarity and depth on each subject, making it a valuable resource for both field professionals and academic researchers alike.

Interestingly, the original intention was not to write a book. Over the years, notes from field experience, combined with valuable contributions from fellow experts, naturally grew into the comprehensive work now before you. I extend my deepest appreciation to all contributing authors for their expertise and insights, and to their organizations for supporting their involvement in this project.

Special thanks to Stan Christman for his wonderful and helpful review of pressure integrity tests; to Rob Crawford for providing valuable information on fiber-optic sensor technologies developed by Well-Sense; to John Perkins from Breakthrough Ventures to provide a cement bond log; and Kirk Harris from Thoroughbond for his comments on cement bond evaluation.

Rahman Ashena

CHAPTER

1

Introduction

Rahman Ashena, Jacqueline Greenwell
Smart Energy Data Analytics (SEDA)-Group, Houston, TX, United States

1.1 Introduction

Well integrity—the ability of a well to contain fluids and gases within designated zones, prevent unwanted migration, and protect both the environment and human health—is fundamental to the safety, efficiency, and environmental responsibility of oil and gas, geothermal (GT), carbon capture, utilization, and storage (CCUS), and other energy transition operations. Ensuring well integrity requires proactive design, thorough monitoring, regular maintenance, and robust risk management throughout the entire lifecycle of a well, from drilling through abandonment. Failure to maintain integrity can result in well control incidents, leaks, environmental damage, and significant economic loss.

With the growing number of wells being drilled and repurposed—especially for energy transition applications—and the extended operational lifespans of many wells, maintaining long-term integrity has become one of the most critical operational priorities for exploration and production (E&P) companies. This is especially true in GT and CCUS operations, where subsurface conditions are significantly more hostile.

In GT systems—whether conventional high-temperature fields or unconventional projects that repurpose inactivated oil and gas wells—the cement barriers must withstand extreme thermal cycling, elevated pressures, and prolonged exposure to chemically reactive fluids. Similarly, CCUS wells often encounter aggressive environments because of high CO_2 concentrations, low pH, and corrosive conditions. These factors can accelerate cement degradation, cause microannuli or channeling, and ultimately threaten zonal isolation and well integrity.

Historically, these harsher and more complex conditions have not received the tailored attention they demand. Cement evaluation has often

Well Cement Integrity
DOI: https://doi.org/10.1016/B978-0-443-51564-4.00002-3

relied on limited tools, such as a single cement bond log (CBL) conducted early in the well's life. However, such isolated assessments are insufficient for high-risk applications. What is needed is a holistic, comprehensive, and transparent methodology for cement evaluation that spans the entire well lifecycle—from cement design to placement, validation, long-term monitoring, and reevaluation. This book fills that critical gap.

For the first time a fully disclosed and systematic cement evaluation methodology is presented in one place—openly shared with both industry practitioners and academic researchers. The methodology integrates laboratory testing, field observations, diagnostic tools, advanced logging techniques, and real-time data analysis into a single cohesive framework. It provides not only technical depth but also practical guidance, making it an invaluable resource for those working in or studying the energy transition space.

The importance of cement integrity in GT and CCUS wells is emphasized throughout this book. Chapters 12, 13, and 14 are specifically dedicated to these sectors, exploring the unique challenges and solutions required in such extreme environments. In addition, GT considerations are addressed across all chapters—including in the Introduction, which outlines the critical role of CBLs and their growing relevance in energy transition wells.

Well-cement integrity is a core element of overall well integrity, and this book positions it at the forefront of operational excellence and innovation in the geoenergy industry. It aims to become a standard reference for engineers, scientists, and decision-makers committed to developing safe, reliable, and sustainable subsurface energy solutions.

This chapter also outlines the scope of the book and its chapter structure, providing a clear roadmap for readers. It highlights the learning outcomes they can expect—ranging from understanding well integrity fundamentals and cement degradation mechanisms to mastering advanced evaluation methods and applying AI-driven analysis in real-world GT and CCUS projects.

1.2 Well integrity and barriers during lifecycle

The life of a field, either conventional GT or petroleum, starts from the exploration phase and ends with its abandonment and final relinquishment. However, the petroleum field life may be extended for longer periods if there are plans for repurposing the field for unconventional GT energy production or CCUS (Fig. 1.1).

The petroleum wells within these fields, until end-of-life, are guided by various internationally industry-accepted standards and practices, including but not limited to the American Petroleum Institute, International

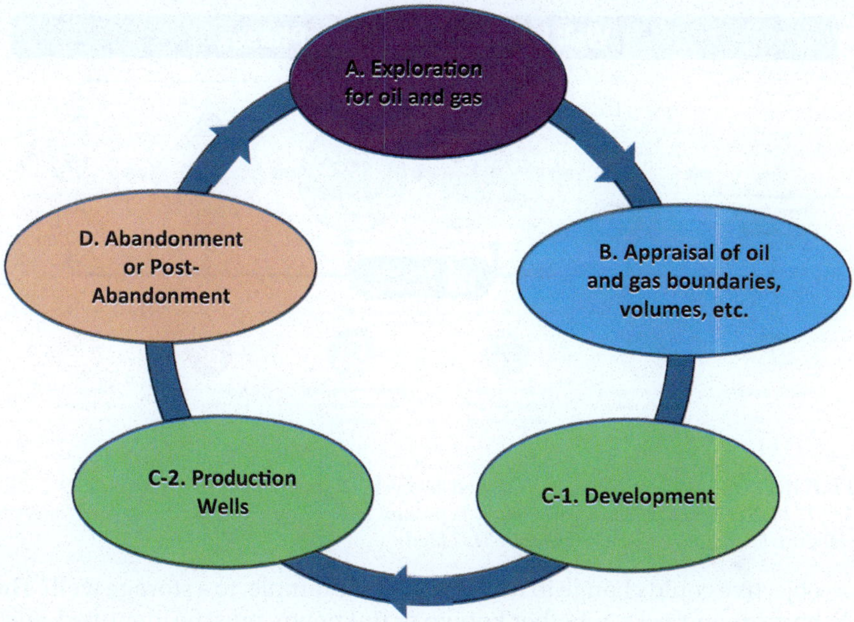

FIGURE 1.1 Different phases of the life of an oil and gas field.

Organization for Standardization (ISO), and Norwegian Shells for Offshore and Kongsberg (NORSOK) standards.

The various elements or components that make up a well are referred to as barriers. A barrier can be defined as *a combination of components or practices that contribute to the well system reliability to prevent or stop uncontrolled fluid flow* (ISO, 2017).

Barrier elements and their ongoing integrity must be designed depending on well conditions, and then verified and maintained throughout the well lifecycle. The evaluation and ongoing validation of cement barrier integrity, whether as a primary or secondary barrier element, is a fundamental responsibility throughout the well lifecycle. Ensuring that cement and all well barriers comply with our industry obligations allows operators to maintain their social license to operate, particularly in an era of increasing scrutiny and rapid media dissemination when issues arise. Refer to Fig. 1.2 showing an outline of a cement barrier lifecycle approach.

Well integrity and lifecycle management are governed by ISO (2017) and are thus divided into the following six distinct phases (Fig. 1.2).

- **Basis of design (BOD) phase** identifies the well objectives, for example, anticipated life span; intended as producer or an injector; will it be fracture stimulated; how will it be completed, for example, free-flow, artificial lift, gas lift, or dual completion; is there a chance the well

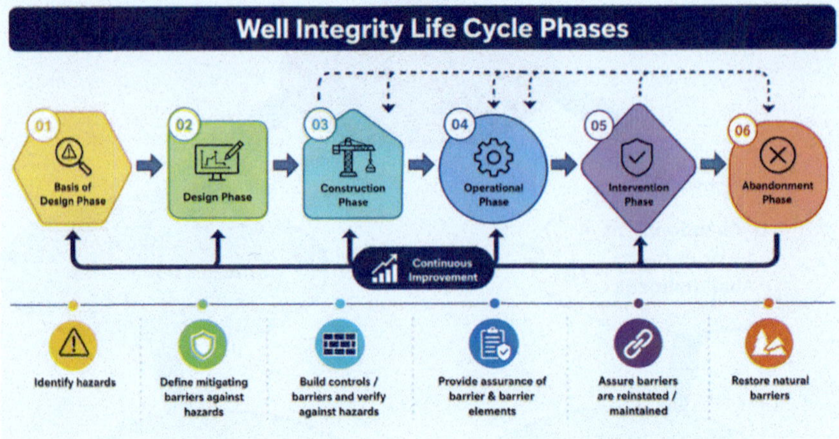

FIGURE 1.2 Well integrity in full well lifecycle *(Redrawn from ISO. (2017). ISO 16530-1:2017:Well integrity lifecycle governance. Petroleum and natural gas industries—Well integrity standard. https://www.iso.org/standard/63192.html).*

objective could change in its lifecycle, for example, to a storage well? The hazards and risks, whether known or unknown, must be identified, such as the anticipated temperatures and pressures, loss zones, subsurface hazards, fluid and/or gas compositions, critical aquifer locations, are they corrosive or freshwater, what are the final abandonment criteria in the jurisdiction, to name a few. A greater number of offset wells in a region improve the understanding of the hazards and risks. Insights from offset well issues should be documented, and continuous improvements for well design should be considered during this phase.

- **Design phase** involves evaluation and selection of the appropriate hardware and processes required to meet the objectives of the BOD. It essentially identifies the controls and adequate number and design specifications of barriers required to minimize risk throughout each well's life and any predicted lifecycle changes. The design phase may include engagement with service providers and specialists. The acceptance and testing criteria to verify barriers must be established in this phase. Cement design and lab testing are critical for long-term cement integrity.
- **Construction phase** involves placing and verifying barrier elements to ensure that they meet design requirements. Any deviations from the activity program, or failures in barrier verification to meet acceptance criteria, must be risk assessed and managed accordingly. Issues with cement blend or placement can result in integrity challenges, potentially impacting productivity or leading to unplanned remediation and increased P&A costs.
- **Operational phase** is typically the longest phase of the well life. Wells must be monitored, maintained, and managed within safe

well operating limits. Any anomalies in well behavior or known integrity breaches should be reported such that they can be diagnosed, risk assessed, and, if warranted, activate a remediation. Ongoing monitoring of cement integrity during this phase is primarily managed by monitoring well annulus pressures and integrity pressure testing.

- **Intervention phase** should include a preassessment of the integrity status of a well before applying the common principles in BOD, design, and construct phases. Barriers are to be revalidated, and any revised integrity status recorded post intervention.
- **Abandonment phase** should include an assessment of the current well integrity status, ensuring that the principles from the BOD, design, and construction phases are applied during the planning and execution of the permanent plug and abandonment (P&A) process. A robust verification of the existing and planned cement integrity is essential to ensure a permanently hydraulically isolated wellbore, preventing the undesired flow of movable hydrocarbons and protecting freshwater aquifers from contamination by hydrocarbons and saline fluids (see Chapter 10 for further information). Any postabandonment failure of cement barriers will result in significantly adverse consequences.

Orphaned wells are an additional industry and regulatory burden. It is the responsibility of a well operator/owner to manage their well portfolio and oversee their associated problems and expenses. However, an orphan well is considered a well that has no identifiable owner, which may be because of a lack of historical records, unexpected bankruptcy, etc. Such wells are usually managed by governmental associations in collaboration with the industry partners, to abandon and reclaim them.

Assessing a well's barrier condition(s) and understanding the associated risks are essential to mitigate or prevent potential adverse consequences. Loss of or compromised barrier integrity could lead to loss of well fluids, hydrocarbons, GT fluids or aquifers, surface or underground well blowouts, crossflow and unwanted fluid migration, loss of reserves or energy sources, additional well component failures, structural collapse, personnel injuries or fatalities, external infrastructure damage or loss, such as a platform or rig, environmental impacts, operational disruptions, financial losses, damage to the reputation of the operator and/or the industry, and unwanted negative media attention or legal proceedings.

Barrier elements are required to perform as intended to maintain integrity and safeguard the total well system reliability. The barriers should be installed as per acceptance criteria, thoroughly tested and evaluated, and monitored, where feasible, using reliable techniques. Well integrity management (WIM) requires an application of techniques with technologies and practices devised to provide a sufficient number of barriers to guarantee integrity in accordance with organizational and regulatory agencies' standards aimed at eliminating or reducing potential risks. In other words, WIM consists of principles, processes, and procedures for planning and implementing suitable barriers and the continued monitoring and maintenance of the same. This barrier philosophy should be applied to each potential flow path and well activity, considering the worst case consequence of a complete loss of well control. Thus operators and service companies must always ensure compliance with the standards.

Barrier elements fall into operational and physical types. An operational barrier is a human action or response that activates a physical barrier such as "kick influx detection" or "pressing the valve to close the blowout preventers (BOPs)." Note that operational barriers by themselves do not constitute a physical barrier. Physical barriers are classified as either hydraulic or mechanical types. Hydraulic barriers consist of hydrostatic fluid pressures preventing loss of well control. Mechanical barriers comprise tangible components or equipment such as BOPs, wellheads, cements, casings, packers, and plugs (see Ashena (2025) for a comprehensive discussion on the list of barrier elements). Since the 1920s, at least two barriers have been required to ascertain the safety of operations during an oil and gas well's lifecycle, which is to ensure that in the event one barrier fails, a leak to the atmosphere does not result; however, in more complicated cases, more barriers may be mandated according to the companies' policies. The two required barriers are briefed as follows:

- *Primary barrier:*
  During drilling the hydrostatic mud pressure overbalance is considered the primary barrier, which is of hydraulic type. During production, primary well barrier elements may include, but are not limited to, subsurface safety valve, production packer, completion string, and the lowest section of casing and cement. See Fig. 1.3; the blue-colored elements are primary (Anders et al., 2015).
- *Secondary barrier:*
  During drilling the BOPs are considered the secondary barrier, which is of a mechanical barrier. During production, such elements include Christmas trees, casing hangers, wellheads, and upper casings and cements. See the mentioned figure above; the red-colored elements are secondary.

Cement plays a critical role in wellbore construction by providing zonal isolation, supporting the casing, and preventing fluid migration between formations. A successful cementing job ensures that the well is properly isolated, secure, and able to withstand the pressures and conditions it will face throughout its lifecycle. Cement is an essential mechanical barrier that is required to seal behind casing strings, both vertically and horizontally, to provide adequate bonding to casing as well as the wellbore or formations.

During drilling's cementing operation, cement slurries act as primary barriers. However, during production and P&A phases, cement sheath acts as both primary and secondary barrier elements. An annular cement

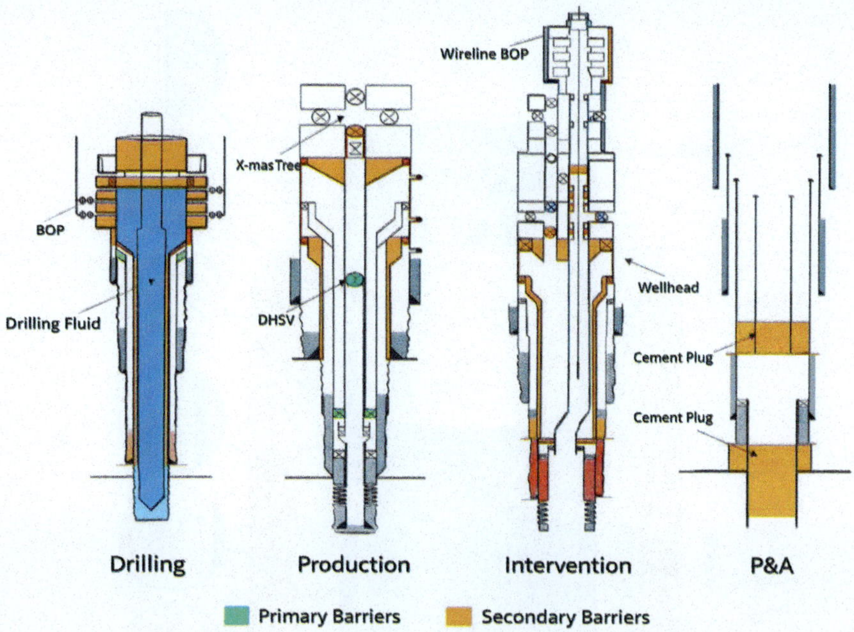

FIGURE 1.3 Principle of two-barrier philosophy during a well's lifecycle for the oil and gas industry. The *blue-colored elements* indicate the primary barriers, whereas the *red-colored elements* comprise the secondary barriers *(Redrawn from Anders, J., Mofley, B., Nicol, S., Vo, N., Weiss, T., & Tore F. (2015). Implementation of well barrier schematic workflows. Paper Presented at the SPE Digital Energy Conference and Exhibition, TheWoodlands, Texas, USA, March 2015. https://doi.org/SPE-173433-MS).*

seal is established between the casing and wellbore; that is, hydraulic isolation of underground formations is created to avoid the development of fluid passageways behind casings between zones of differing pressure regimes. In the event of drilling hole problems such as lost circulation or for P&A, cement plugs may be set in the wellbore to seal the bottom and top sections of the well. Fig. 1.4 illustrates cement failure causing hydraulic communication, fluid flow, or gas migration.

There are several blowout case studies worldwide that occurred because of lack of or complete failure of cement integrity, the most famous of which was probably the Macondo Deepwater Horizon in 2010. The BP's report on the incident published in March 2011, titled "Investigation of the Macondo Well Blowout Disaster," mentioned that the first two reasons for the blowout were "the annulus cement barrier did not isolate the hydrocarbons" and "the shoe track barriers did not isolate the hydrocarbons," and that a misconduct or wrong decision during well completion that led to the blowout was "not to run cement bond log (CBL)" on the basis of cement lift pressures and the absence of fluid losses during cementing (see Fig. 1.5 for the well schematic). It is interesting that though the Schlumberger crew

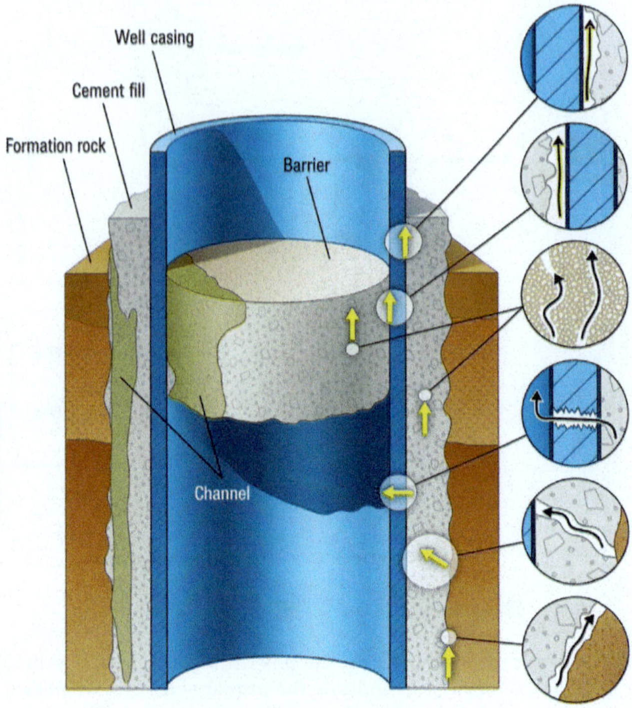

FIGURE 1.4 The relation of cement bond and well integrity shown schematically. *(Redrawn from Offshore-Map. (2025). Offshore Print Articles. Volume 77-05. Retrieved January 6, 2025, from https://img.offshore-mag.com/files/base/ebm/os/image/2017/05/content_dam_offshore_print_articles_volume_77_05_1705offslb_p01.png?auto=format,compress&fit=max&q=45&w=950&width=950).*

had arrived at the Deepwater Horizon to perform cement bond logging, BP concluded that CBL was not required, and thus the crew was released and departed from the rig without running the log (BP, 2010). Following the incident, cement bond logging is not only a recommended practice, but it is also usually standard in offshore operations. Compliance depends on the location, jurisdiction, and regulations. Local regulatory bodies in the United States or elsewhere may impose additional requirements or mandatory measures for cement evaluation. In the United States [such as the Bureau of Safety and Environmental Enforcement (BSEE)], regulations often require CBLs, particularly in offshore operations, to ensure well integrity and prevent the migration of fluids or gases.

1.3 Cement integrity in different industry sectors

Well-cement integrity is of great importance in various industry sectors to guarantee the safety of well operations technically and environmentally

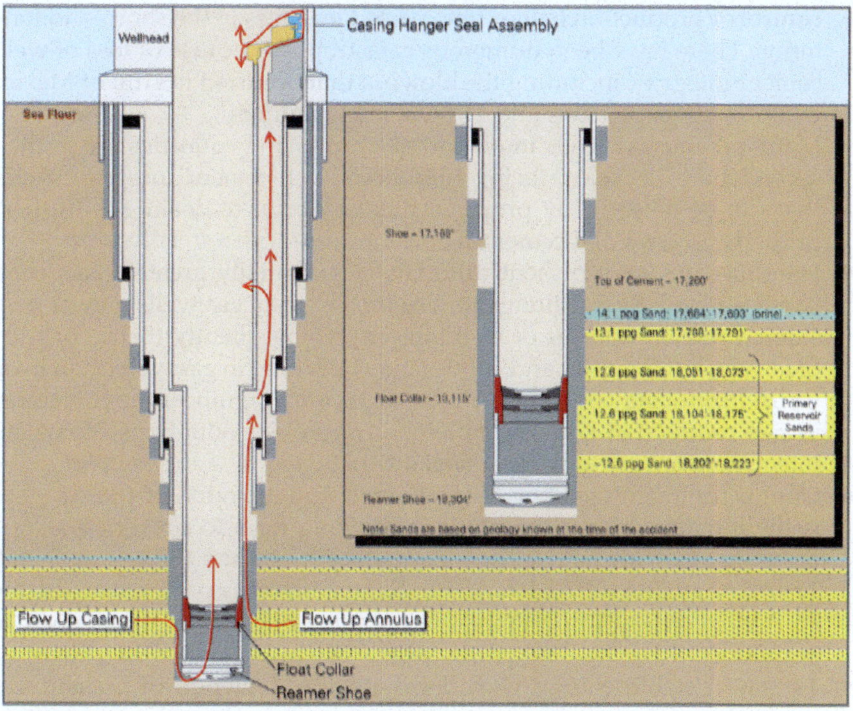

FIGURE 1.5 Possible flow path through the cement leading to the Deepwater Horizon's Macondo Blowout *(Redrawn from BP. (2010). Deepwater horizon accident investigation report).*

for the short- and long terms. These sectors consist of energy industries such as GT, CCUS, and conventional and high-pressure high-temperature (HPHT) oil and gas industries.

The importance of cement integrity in these sectors is elaborated with some facts and figures as follows:

- Global oil reserves were estimated to be from 1565 to 1730 billion barrels as of December 2022, based on different sources. According to International Energy Agency (IEA) approximations, the current global demand for oil in 2023 is about 101 million barrels per day, and it will increase by another 880,000 barrels per day in 2024. As for natural gas, global reserves were 7257 trillion cubic feet based on the US Energy Information Administration (EIA) as of January 2020, whereas the global gas demand is 362 billion cubic feet per day, which would almost stay stable in 2024. Considering the reserves and demands, oil and gas markets have been well established for many years and will continue to be stable for at least the next 50–60 years. Based on most oil and gas producing countries' national visions and aims to stay sustainable top energy suppliers globally, it is vital to prevent any risks of loss of well control or production halt or unexpected leakages in the short- and long terms. There have been numerous catastrophes because of loss of well-cement integrity, including the blowout that occurred in Gulf of Mexico in 2010 with explosions and 11 personnel fatalities.

- Many oil and gas wells in oil and gas producing countries are in their second half-lifecycles, facing numerous well-cement integrity issues that can be detected by proper evaluation prior to dealing with them properly by remedial cementing.
- Handling gas may present different or potentially greater risks compared with oil; thus drilling and production from gas wells may be considered riskier in terms of well integrity. Consequently the probability of losing well integrity and control may be higher in gas wells than in oil wells. Producing countries or companies with significant involvement in gas production should implement rigorous evaluations of cement integrity to ensure long-term well integrity.
- One of the main challenges of plugged and abandoned (P&A) wells is the integrity of cement plugs in the long term to prevent leakages or pathways with fluid intrusions to the surface and the environment. Therefore evaluating and verifying cement integrity prior to and during P&A is of critical importance to guarantee integrity is maintained in perpetuity.
- To stay a stable energy provider, with the depletion of oil and gas reservoirs, it is necessary to exploit deeper hydrocarbon resources, which means drilling and producing from HPHT wells. Because of high pressures and temperatures and possibly sour gases, the evaluation of cement integrity in such wells is critical to maintain an intact barrier.
- Some countries rightly identified reaching net-zero carbon construction by 2050 as a priority goal and, to this end, are urgently seeking innovative ways to reduce carbon emissions. Having said that, CCUS plans are going on with an increasing trend in drilling CCUS wells. However, such wells may encounter integrity issues because of high corrosion rate and possible chemical degradation of cement exposed to injected corrosive CO_2 gases. Cement integrity in such wells should never be compromised, but rather, following proper planning and execution, it should be strictly evaluated using the best possible methods to guarantee safe field operations.
- Diversification of energy production using renewables is also one of most countries' long-term energy targets. The GT energy market is a potential segment of the energy market proliferating because of the climate goals being opted for by different nations. Based on IEA reports, GT energy market and demand are growing with global attention to and investment in renewables. In 2021 27 countries, including the United States, generated about 92 billion kWh of electricity from GT energy (based on EIA), which is equivalent to about ten 1000-mW power

plants, showing how the industry is in its infancy. Well-cement integrity in forthcoming GT wells, particularly abandoned wells, is of critical concern. As temperature and brine salinity are considerably greater in conventional GT wells than petroleum wells, they can disintegrate the cement more rapidly. In addition, as the margin of profit in GT is narrower than the oil and gas sector, preventing integrity failure and associated costs is of greater importance to maintain favorable economic. Therefore cement integrity should be paid attention to the fullest in such wells, and its evaluation should be strictly handled prior to production.

1.4 Scope of well-cement integrity and evaluation

Well-cement integrity is not a new topic; its evaluation started in the 1930s with an estimation of top of cement depth, using temperature logs. In the 1960s it continued with acoustic cement bond logging tools, and since then the tools have been developed chronologically. Since then, several other methods have been developed and applied in the field. Therefore the important features of the book consist of the following:

- ✓ Experimental lab cement integrity evaluation;
- ✓ Sustained casing pressures' (SCPs') monitoring, prevention, and remediation;
- ✓ CBLs sonic or ultrasonic with their evaluation;
- ✓ Positive pressure tests (PPTs), leak-off test (LOT), and formation integrity test (FIT);
- ✓ Inflow or negative pressure test (NPT);
- ✓ Temperature and noise logs;
- ✓ Fiber optic logs;
- ✓ Corrosion logs;
- ✓ Cement plugs evaluation;
- ✓ AI/ML data-driven cement evaluation;
- ✓ Holistic cement evaluation;
- ✓ P&A and well-cement integrity; and
- ✓ Cement integrity in HPHT, GT, and CCS wells.

Therefore this book will discuss well-cement integrity with a focus on evaluation fundamentals in 13 chapters. Following the introduction about well-cement integrity and its importance in different industry sectors in Chapters 1 and 2, it discusses the holistic cement evaluation, which is essentially a holistic overview of cement evaluation by discussing all the essential evaluation methods and considerations made for a comprehensive cement barrier integrity assessment.

Chapter 3 discusses experimental cement evaluation, including bond tests conducted within a lab environment to assess cement integrity.

Chapter 4 covers the SCP topic and its monitoring role in the detection of loss of barrier integrity. It also discusses prevention and remedial measures to such events.

Chapter 5 covers the concepts of CBLs and tools, which are widely used for investigating the cement bond using acoustic and ultrasonic waves. Next, wellbore conditions influencing CBL data are covered, including pressure, temperature, parameters adjustment, eccentricity, cement sheath, compressive strength, pipe diameter and thickness, mud properties and cake, lithology, hole shape and well deviation, and the microannulus effect and its possible misleading effect on CBLs' interpretation, followed by a practical method to eliminate the said effect. Readers will become familiar with the standard method of applying surface pressure to eliminate the mentioned effect. Next the chapter delivers a practical method for verifying CBL quality or reliability control using repeat sections and variations in transit time curve. The last section of the chapter delivers a recommended operational procedure with guidelines for supervising and witnessing cement bond logging operations to ensure good practice and reliability of measurements taken. The last part of the chapter presents some practical case studies with detailed interpretations/explanations, for both cases of sonic and ultrasonic logs.

Chapter 6 discusses the pressure integrity tests (PITs), consisting of conventional PPT, LOT, FIT, and inflow tests. The inflow test, or the NPT, is practically explained to verify well-cement integrity in shoe tracks and liner laps or across multiple casing pipes/overlaps where PPTs do not either apply or their findings are not reliable enough.

Chapter 7 covers temperature and noise logs and fiber optic logs, which are emerging technologies to complement well integrity evaluation as effective evaluation tools for flow detection behind pipes, which can be considered crucial cement integrity evaluation techniques. Tracer logs are also briefly mentioned because they are not used nowadays.

Chapter 8 discusses corrosion logs, which can indirectly hint about cement conditions behind the casing.

Chapter 9 focuses on evaluating cement plugs to assess their integrity and ensure effective wellbore isolation. This chapter is the prerequisite of Chapter 13.

Chapter 10 addresses data analytical AI/ML approaches to cement bond evaluation using a history of previous cementing and cement bond data.

Chapter 11 covers the P&A fundamentals and methods with some field case studies. The prerequisites to this chapter, including Chapter 9, were already covered.

Chapter 12 focuses on cement integrity in CCUS wells, exploring the unique challenges and requirements for ensuring effective wellbore sealing in CCUS applications. It highlights the importance of cement performance in maintaining long-term well integrity and preventing leaks in the context of CO_2 storage.

Chapter 13 discusses cement bond evaluation in HPHT and GT wells, including consideration of the effect of temperature on the log readings and tools.

This book provides readers with a thorough understanding of the theories and methodologies, along with practical approaches, to effectively apply comprehensive well-cement integrity evaluation measures. Because of its alignment to global safety priorities for safeguarding well integrity, this work will have an interconnected environmental, socioeconomic, and scientific relevance worldwide.

1.5 Learning outcomes

There are multiple key learning outcomes within this work, categorized into four classes. At the conclusion of the book, readers should be able to understand and apply integrity evaluation methods of annular cements and cement plugs. Specifically, they are expected to learn how to apply an overall holistic approach for a comprehensive cement bond evaluation, comprising the following:

1. *Experimental lab cement integrity evaluation:*
   a. Learn fundamentals of cement slurry, bond, and integrity evaluation.
2. *SCP:*
   a. Learn about SCPs, their prevention, monitoring, and controlling/cure methods.
3. *Annular cement integrity:*
   a. Learn fundamentals and concepts of CBLs and tools and detect the effect of wellbore conditions on CBL readings,
   b. Understand reasons for microannulus creation at cement-casing and cement-formation interfaces, its misleading effect on interpretation, and the practical method of eliminating its effect, and how to differentiate microannulus from microchannels,
   c. Learn the guidelines that should be followed for quality or reliability control of CBLs,
   d. Learn to apply a standard method of supervising and witnessing cement bond logging operations to ensure reliability of logs during acquisition,
   e. Learn how to interpret sonic and ultrasonic CBLs for cement integrity evaluation using bond index and other criteria and rationalize their readings based on the conditions of real cases,

f. Learn about PITs, including the conventional PPT, LOT, FIT, inflow test/NPT for channeling detection from shoe track, liner laps, and cement plugs,
g. Learn about temperature noise logs, fiber optic logs, and tracer methods for cement evaluation after cementing, with some related case studies,
h. Learn about corrosion logs for direct casing evaluation and indirect cement evaluation, with field case studies.

4. *Data analytics in cement evaluation:*
   a. Learn about data analytical and AI/ML-driven methods of cement evaluation.
5. *Cement plugs and P&A:*
   a. Learn about different cement plugs with well integrity and evaluation methods,
   b. Understand principles of P&A and well-cement integrity evaluation.
6. *Industry-specific wells:*
   a. Learn about cement integrity in the corrosive environments of carbon capture utilization and storage (CCUS) well and
   b. Learn about cement integrity and evaluation in high-temperature (HPHT) wells and GT wells.

References

Anders, J., Mofley, B., Nicol, S., Vo, N., Weiss, T., & Tore F. (2015). Implementation of well barrier schematic workflows. Paper Presented at the SPE Digital Energy Conference and Exhibition, The Woodlands, Texas, USA, March 2015. https://doi.org/SPE-173433-MS.

Ashena, R. (2025). *The practical handbook of well control*. Routledge, Taylor and Francis Group https://www.routledge.com/The-Practical-Handbook-of-Well-Control/Ashena/p/book/9781032753850?srsltid=AfmBOooZqOJyBoNQ8lWhWyMcNMmdX4-St1TUa2goxJzlN_6-31cO4lBo.

BP. (2010). Deepwater horizon accident investigation report.

ISO. (2017). ISO 16530-1:2017: Well integrity lifecycle governance. Petroleum and natural gas industries—Well integrity standard. https://www.iso.org/standard/63192.html.

CHAPTER

2

Holistic cement evaluation

Rahman Ashena, Jacqueline Greenwell
Smart Energy Data Analytics (SEDA)-Group, Houston, TX, United States

2.1 Introduction

Cement barriers, like most well barriers, are subject to degradation over time and exposure to subsurface conditions. A commonly performed evaluation, such as a cement bond log (CBL), typically occurs early in the well's life—usually postdrilling and before completion activities. However, the durability of the cement barrier should not be defined by a single assessment at one point in time. The effectiveness of the cement as a well barrier must be evaluated throughout the entire lifecycle of the well, from repurposing to postabandonment.

Holistic cement evaluation is a comprehensive and integrated approach that leverages all available evaluation techniques and observations to ensure cement integrity, zonal isolation, and long-term well stability. This approach goes beyond basic cement placement and evaluation, encompassing every phase of the cementing process—from initial design through to postplacement evaluation. By integrating experimental lab tests, field cementing observations, diagnostic tests, and various logs, a thorough assessment of the cement job is made possible. This helps assess both the immediate quality of the cement placement and its long-term performance.

By combining various sources of data and providing a breadth of knowledge on the downhole characteristics of a well, such as lithology, pressures, hole conditions, cementing data, cement logs, temperature data, noise logs, corrosion or multifinger caliper logs, and fluid or gas analysis, we can begin to "view" the downhole integrity with ever-increasing accuracy. Ultimately this cohesion of information will lead to effective decision-making, optimized productivity, cost savings, successful operations, and reduced risk.

The holistic approach is especially necessary for more hostile environments such as geothermal and CCUS cases, where extreme temperatures,

Well Cement Integrity
DOI: https://doi.org/10.1016/B978-0-443-51564-4.00001-1

pressures, and chemical exposures can significantly accelerate cement degradation and challenge long-term integrity.

2.2 Cement integrity lifecycle review and evaluation process

The following is the suggested data that should be compiled and analyzed, whether examined by competent engineers or by future artificial intelligence/machine learning (AI/ML) data analytical tools, to establish and understand the level of cement barrier integrity/competence, the limitations, and the potential risks associated with the in situ cement barrier(s).

Recommended data to review are, where available, but not limited to (see Fig. 2.1):

- ✓ Review well construction and intervention programs and objectives, including cement labs' tests.
- ✓ Review well trajectory, deviation, dogleg severity, and hole conditions, which could affect cement integrity evaluation.
- ✓ Review hazards identified and risks determined using offset wells for cementing prior to well construction and compare the effectiveness of the identified mitigations and controls that were implemented.
- ✓ Confirm zonal isolations to be achieved in line with regulations and standards.
- ✓ Review initial acceptance criteria against regulations and standards and any deviations during cementing operations.
- ✓ Review the casing running report and the daily cementing operations report. Note depths/formations encountering hole problems, cavings, faults, washouts, fill, drilling losses, stuck pipe issues, influx/kick, etc.
- ✓ Review geological reports to understand the impact of lithology and wellbore conditions on CBL responses.
- ✓ Review open-hole caliper log data for integrated hole diameter and volume useful for cementing, where available. Review actual bottom-hole temperature and pressure versus estimated values.
- ✓ Review planned/expected (depths of top of) cement (TOC) versus actual or calculated.
- ✓ Review records of cement fill (tag) inside casing postcement placement.
- ✓ Review pressure integrity tests (positive pressure test, PPT, and negative pressure test, NPT, or inflow test) before drill-out of casing shoe cement.
- ✓ Review premature slurry returns to surface, losses during cementing, plug bump, pressure tests, and bleed back volumes.
- ✓ Review slump and surface top-up volumes.
- ✓ Review cementing job procedure and service provider pumping charts; effective hole cleaning and pump rates; any operational issues during

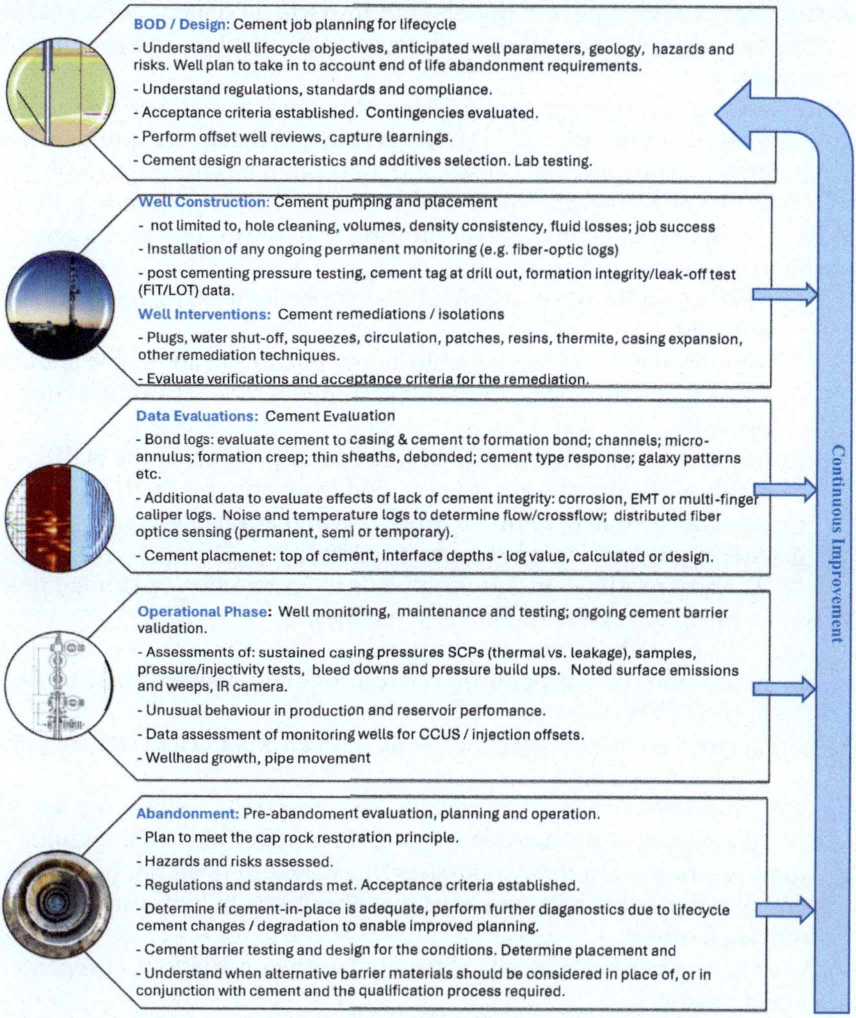

FIGURE 2.1 Different phases of well lifecycle and holistic well-cement integrity approach.

pumping, slurry density issues, surface sample curing times, compressive strengths, etc.

✓ Review formation integrity tests (FITs) and leak-off tests (LOTs) post (casing) shoe cement drill out.

✓ Review any possible temperature noise logs run within the cement setting time window.

✓ Review cased-hole cement evaluation log(s), where available.

✓ Highlight potential risks for noncompliant cross-flow or inadequate barrier competency or lack of barrier height of competent cement.

- ✓ Confirm cement type and understand impacts on cement bond evaluation, for example, foamed cement, lightweight cement, and expandable cement.
- ✓ Review log quality control, logging parameters, and tool calibrations.
- ✓ Interpret potential risks such as channeling and the microannulus effect. Understand the potential causes and associated hazard impacts, if any.
- ✓ Utilize the expertise of cement interpretation analysts when required.
- ✓ Review operational lifecycle monitoring to confirm cement integrity or failure of same;
  - ○ Monitor annulus for sustained casing pressure (SCP), not thermally induced.
  - ○ Monitor if SCP is maintainable below maximum allowable annular surface pressures (MAASP)/maximum operational annular surface pressure.
  - ○ Monitor bleed-down and pressure buildup trends of the SCP.
  - ○ Analyze results of samples obtained from annulus SCP.
  - ○ Analyze results of annulus fluid tops and applied PPTs.
  - ○ Analyze results of inflow pressure tests.
  - ○ Analyze results of monitoring wells in an injection or storage field.
- ✓ Recommend or review (if data is available):
  - ○ CBLs are either sonic or ultrasonic.
  - ○ Noise and/or temperature-related logging; conventional or fiber optic (DFOS) logs.
  - ○ Corrosion logs that may indirectly indicate insufficient cement bond protecting casing(s).
  - ○ Cross-reference any applicable log response within open-hole, cased-hole, or production logging data with CBL, lithology, etc.
- ✓ Analyze production data anomalies that relate to potential breaches of cement barriers connecting zones that should be in hydraulic isolation from each other.
- ✓ Analyze any verification data associated with remediations or repairs to cement barriers.
- ✓ Review cement plug tests and verification.
- ✓ Review if the cement barriers required by company policies or regulations for cap rock restoration have been met for plug and abandonment.

This list is by no means exhaustive but provides a guide, along with the details provided in the chapters of this book, to provide a well integrity engineer with a sound understanding and the capability to analyze cement as a barrier throughout the well lifecycle. This combined data analysis approach may be utilized in future AI/ML software, as technology advances, to provide both quantifiable and qualitative assurances as to the cement integrity and risk profile throughout the well life and longevity post well abandonment.

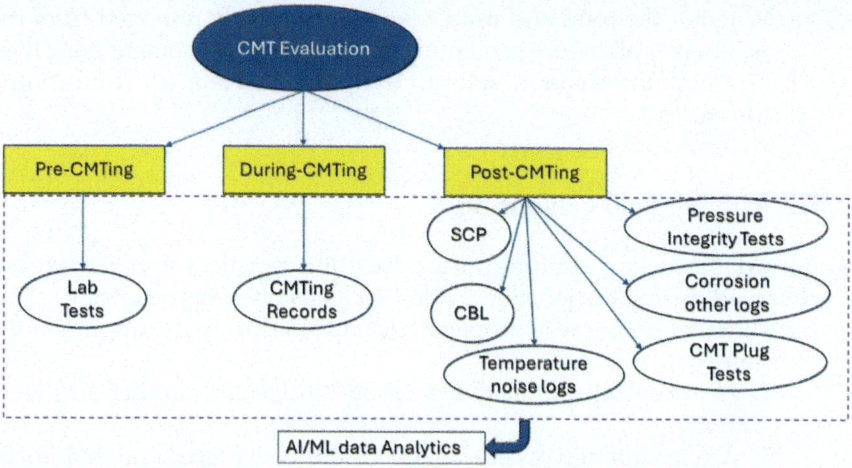

FIGURE 2.2 Elements of Holistic cement integrity evaluation. *AI/ML*, Artificial intelligence/machine learning; *CBL*, cement bond log; *SCP*, sustained casing pressure.

2.3 Main evaluation elements

In the holistic cement evaluation, several evaluation elements should be considered, which are categorized as follows (see Fig. 2.2):

2.3.1 Precementing

1. Cement experimental/lab testing: including cement slurry tests such as thixotropy test as well as cement mechanic tests, particularly the shear-bond test and push-out test (the hydraulic bond). Refer to Chapter 3 for more information.
2. Analysis of cement job design: Ensure that the cement and cementing job design features, including the type of cement slurry and the method of placement, were tailored to the specific well construction design (such as casings, well conditions, and its challenges).
   a. Whether an appropriate type of cement and additives as well as the spacer fluid were selected,
   b. Whether correct cement density was estimated,
   c. Whether the correct cement volume required was determined based on caliper logs or sound safety factors.
   d. Whether the rathole (the spacing between bottom-hole and casing shoe) was short and if the rathole was filled with mud heavier than the cement density.

  - e. Whether the mud and mud cake were properly removed from the wellbore wall before cementing to ensure proper cement bonding,
  - f. For more information, refer to Chapter 4, Section 4.6 "Controlling Measures."

2.3.2 During field cementing

3. Record the casing running and cementing practices and indications observed during the job. Some items to focus on are as follows:
   - a. Whether casing was properly centralized during its running in the hole,
     - i. Were centralizers installed based on design (number and location)?
     - ii. Was installation of centralizers carefully executed and monitored by the driller or the mud-man or operator company's representative (depending on company's policy)?
   - b. Whether proper spacer fluid (with good properties) was used between the mud and cement slurry to prevent cement contamination and allow mud and cake removal,
   - c. Whether cementing was conducted with compliance to the cement plan,
   - d. Whether safety meeting was held,
   - e. Were the cementing lines pressure tested? Were they recorded?
   - f. Whether the cement was mixed properly (during pumping),
   - g. Whether the correct equipment was used,
   - h. Whether wiper plugs were used for separating the cement slurry from spacer fluid?
   - i. Whether high enough displacement rate was maintained during the displacement process to remove mud and mud cakes off the wellbore wall,
   - j. Whether the right amount/volume of cement slurry with right cement density was pumped,
   - k. Whether there were no indications during the cementing job implying deviations from the planned successful placement, such as slurry losses, formation fluid kick, density fluctuations, ineffective hole cleaning, pressure pump pressure anomalies, or problems,
   - l. Whether surface observations were made during cementing such as
     - i. shearing,
     - ii. bumping,
     - iii. cement return was observed at the surface (if company policy requires that)?
   - m. Whether enough wait on cement time has passed?

n. Whether TOC was found (during cement drill out) near the design depth,
o. Whether any annular pressures were recorded after cementing.

2.3.3 Postcementing

4. *SCP* monitoring provides SCP values in different annuli. These values should be checked with MAASP values, using bleed-down and buildup behaviors. A sample analysis may be provided. During production or injection, monitoring wellbore annular pressures can reveal if there is any hydraulic communication between different formations because of cement failure or incomplete cementing. Refer to Chapter 4 for more information.
5. *CBLs (sonic or ultrasonic)* are the traditional method for assessing cement bond quality, and their interpretation typically requires analysis by experienced experts. This technique uses acoustic or ultrasonic waves to assess the quality of the cement bond to the casing and the formation. A good bond is inferred when significant signal attenuation is observed, indicating strong cement-to-formation and cement-to-pipe bonding. However, a bad bond is inferred when minimal acoustic signal loss is observed, suggesting poor cement placement or damage, potentially leading to insufficient zonal isolation and compromised well integrity. Refer to Chapter 5 for more information.
6. *Pressure integrity tests* include the LOT and FIT, which are conducted during the drilling phase, as well as the PPT and the inflow test or NPT, which are performed after cementing to ensure wellbore integrity and pressure control. Refer to Chapter 6 for more information.
7. *Temperature and noise logs*, conducted after cementing, which can record noise and temperature parameters in real time over the entire length of the well simultaneously, help pinpoint the depths of fluid and gas movement behind the pipe by detecting temperature anomalies and minor noises. These logs, therefore, contribute to identifying leakage areas of incomplete circumferential cement isolation. Refer to Chapter 7 for more information.
8. *Fiber optic logs*, deployed as permanent, semipermanent, or temporary systems, which may be considered modern temperature noise logs, provide real-time monitoring of parameters such as strain and temperature along the entire length of the well. This new technology is increasingly used for accurate cement integrity evaluation and identification of potential issues such as inadequate bonding or microannuli. Refer to Chapter 7 for more information.
9. *Corrosion logs* provide data that can be used to assess the extent of external casing degradation, which may indicate areas of inadequate cement

coverage and highlight potential zones of poor cement bonding. Refer to Chapter 8 for more information.

10. *Cement plug tests* are conducted to assess the integrity and effectiveness of the plug in providing proper wellbore isolation and preventing fluid migration or ensuring the long-term stability of the well. Refer to Chapter 9 for more information.
11. *AI/ML methods* provide an innovative potential method for a data-driven cement evaluation owing to AI/ML' capability to harness data and provide more accurate, efficient assessments. The success of these technologies, however, hinges on the quality and volume of the input cementing and cement evaluation data. Regions with abundant good-quality data from a high volume of wells are well-positioned to leverage AI and ML for enhanced cement performance analysis because good training can be achieved. However, this method may not be a proper method in the case of low-volume or low-accuracy data or when there are uncertainties about the data; for example, wells often change operators over their lifespan, creating challenges such as poorly managed or lost data. In such cases, expertise and competency are essential to ensure the accuracy and reliability of integrity evaluations, particularly when addressing regional complexities and industry-specific challenges. Refer to Chapter 10 for more information.

2.4 Summary

Holistic cement evaluation ensures that cementing operations are effective, reliable, and capable of providing long-term wellbore stability. By integrating laboratory testing, real-time field cement job data, advanced logging techniques, and pressure testing, operators can ensure that their cement job delivers cement integrity by providing zonal isolation and preventing fluid migration for its entire lifecycle. To prevent severe well integrity issues, a comprehensive approach allows for timely evaluation, preserving the integrity and safety of the well. This chapter provides a comprehensive checklist as well as elements for holistic cement evaluation, covering all three phases: precementing, during cementing, and postcementing. These elements will be covered in the next chapters.

CHAPTER

3

Experimental lab cement evaluation

Jocin J. Abraham[a], *Catalin Teodoriu*[b], *and Mahmood Amani*[a]

[a]**Texas A&M University at Qatar, Doha, Qatar,** [b]**University of Oklahoma, Norman, OK, United States**

3.1 Introduction

In any oil and gas drilling operation, laboratory testing of the cement to be used is a critical step that must be completed before beginning the drilling and cementing process. This ensures the proper formulation is selected, tailored to the specific conditions both at the surface and downhole, and in alignment with industry best practices. It is essential to test the actual on-site materials and the cementing additives to evaluate their properties, as there may be variations between different cement batches. This thorough testing helps guarantee optimal performance and reliability throughout the operation.

Nearly all oil well cement testing standards and laboratory procedures are defined by the American Petroleum Institute (API), which has established a set of guidelines to be followed before any cementing operation. Although these API standards form the foundation, they do not encompass the full spectrum of available tests today. To address this, additional standards from organizations, such as the International Standards Organization (ISO) and the American Society for Testing and Materials (ASTM), are often used in conjunction with the API guidelines. This combined approach ensures consistent testing, upholds industry standards, and helps eliminate discrepancies during both laboratory testing and actual cementing operations.

Well Cement Integrity
DOI: https://doi.org/10.1016/B978-0-443-51564-4.00004-7

Among the API standards, cement testing broadly falls under the API-10A and API-10B standards, and the most common specifications are listed below (API, 2024):

- API Specification 10A: "Cements and Materials for Well Cementing"
- API RP 10B-2: "Recommended Practice for Testing Well Cements"
- API RP 10B-3 (ISO 10426-3): "Testing of Well Cements used in Deepwater Well Construction"
- API RP 10B-4 (ISO 10426-4): "Preparation and Testing of Foamed Cement Formulations at Atmospheric Pressures"
- API RP 10B-5 (ISO 10426-5): "Recommended Practice on Determination of Shrinkage and Expansion of Well Cement Formulations at Atmospheric Pressure"
- API RP 10B-6 (ISO 10426-6): "Recommended Practice on Determining the Static Gel Strength of Cement Formulations"

API Specification 10A is the industry standard that outlines the specifications and guidelines for preparing "neat" samples of the six different classes of cement—Class A, B, C, D, G, and H. It also defines the physical, mechanical, and chemical composition requirements for each class of cement (API, 2019). Table 3.1 presents a simplified version of the chemical requirements for the different cement classes as recommended by API. The API 10A also recommends guidelines for several cement tests for the neat cement samples without additives—including those for compressive strength, free fluid test, consistency/fineness, thickening time, and others. For cements with additives and for additional recommended specifications, API 10B was introduced to cover a more comprehensive set of tests for different scenarios and for cement slurries with different additives in the formulation. Several of these recommended API tests are discussed in detail below. Though these tests are not comprehensive, their main purpose is to create a minimum required standard for oil well cement used in the field to guarantee a certain degree of safety and reliability.

This chapter discusses various laboratory tests used to evaluate the properties of cement slurries and cement sheaths, which are crucial for assessing well cement integrity. The slurry tests include cement density, fineness test, and consistency (or thickening time) test, all of which help determine the fundamental characteristics of the cement slurry; rheology tests are highlighted to evaluate the flow behavior of the cement slurry; and the free-water and fluid loss tests are critical for understanding the slurry's stability and filtration characteristics. As for cement sheath, the permeability test provides insights into the potential for fluid migration, and the thermal conductivity test is essential for understanding the heat transfer properties of cement sheath. Next, both destructive and nondestructive compressive strength tests are conducted to evaluate the mechanical strength properties of the cement. Furthermore cement bond strength

TABLE 3.1 API recommended compositions for different cement classes, Adapted from API (2019).

	Cement class					
	A	**B**	**C**	**D**	**G**	**H**
Ordinary grade (O)						
Magnesium oxide (MgO), maximum (%)	6	NA	6	NA	NA	NA
Sulfur trioxide (SO_3), maximum (%)						
Loss on ignition, maximum (%)	3.5	NA	3.5	NA	NA	NA
Insoluble residue, maximum (%)	1.5	NA	1.5	NA	NA	NA
Tricalcium aluminate (C_3A), maximum (%)	NR	NA	15	NA	NA	NA

(continued on next page)

TABLE 3.1 API recommended compositions for different cement classes, Adapted from API (2019)—cont'd

	Cement class					
	A	B	C	D	G	H
Moderate sulfate-resistant grade						
Magnesium oxide (MgO), maximum (%)	NA	6	6	6	6	6
Sulfur trioxide (SO_3), maximum (%)	NA	3	3.5	3	3	3
Loss on ignition, maximum (%)	NA	3	3.5c	3	3	3
Insoluble residue, maximum (%)	NA	0.75	1.5	0.75	0.75	0.75
Tricalcium silicate (C_3S) (%)	NA	NA	NA	NA	48–58	48–58
Tricalcium aluminate (C_3A), maximum percent	NA	8	8	8	8	8
Total alkali content, expressed as sodium oxide (Na O) equivalent, maximum (%)	NA	NR	NR	NR	0.75	0.75

(continued on next page)

High sulfate-resistant grade						
Magnesium oxide (MgO), maximum (%)	NA	6	6	6	6	6
Sulfur trioxide (SO_3), maximum (%)	NA	3	3.5	3	3	3
Loss on ignition, maximum (%)	NA	3	3.5	3	3	3
Insoluble residue, maximum (%)	NA	0.75	1.5	0.75	0.75	0.75
Tricalcium silicate (C_3S) (%)	NA	NR	NR	NR	48–65	48–65
Tricalcium aluminate (C_3A), maximum (%)	NA	3	3	3	3	3
Tetracalcium aluminoferrite (C_4AF) plus twice the tricalcium aluminate (C_3A), maximum (%)	NA	24	24	24	24	24
Total alkali content expressed as sodium oxide (Na_2O) equivalent, maximum (%)	NA	NR	NR	NR	0.75	0.75

NA, *not applicable;* NR, *no requirement*.

FIGURE 3.1 (A) Cement mixer, Adapted from Abraham et al. (2025) and (B) weighing scale, Adapted from Abraham et al. (2024).

tests are discussed, including shear-bond tests and push-out or hydraulic bond strength tests, essential for assessing the quality of the cement sheath and its ability to ensure long-term well integrity.

3.2 Slurry preparation and sampling

API Specifications 10A and 10B provide comprehensive guidelines on cement slurry formulation and preparation procedures, outlining the necessary steps for all cement slurry testing in accordance with API standards. Equipment used for slurry preparation includes weighing balances, sieves, and a mixing device—which is typically a bottom-drive, blade cement mixer—as shown in Fig. 3.1. The mixing device should be capable of mixing cement slurry at speeds of 4000 ± 250 and 12,000 ± 500 RPM.

Cement needs to be sieved before slurry preparation to ensure that the cement is a fine powder and that there are no clumps or agglomerations in the cement. Distilled or deionized water should be used for preparing the cement slurry, and the temperature of both the cement and water prior to mixing should be at room temperature, approximately 23°C. Table 3.2 provides the mixing quantities of dry cement and water as per API specifications for different classes of cement.

For slurry preparation, the mixing container, with the required mass of water for the class of cement to be prepared, is placed on the mixer base.

TABLE 3.2 Mixing quantities for cement and water for different classes of cement (in g), Adapted from API (2019).

Components	Class A and B (g)	Class C (g)	Class D and H (g)	Class G (g)
Mix water	355 ± 0.5	383 ± 0.5	327 ± 0.5	349 ± 0.5
Cement	772 ± 0.5	684 ± 0.5	860 ± 0.5	792 ± 0.5

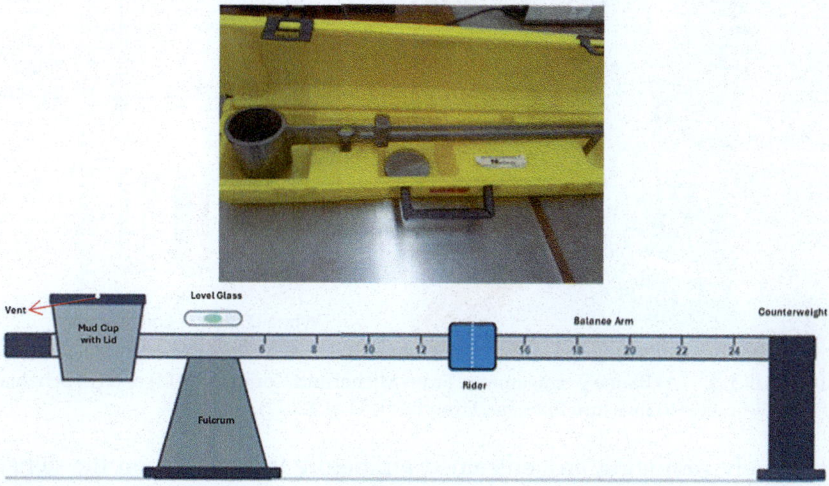

FIGURE 3.2 (A) Pressurized fluid density balance for density measurements (*Redrawn from Igbani, S., Peletiri, S. P., & Egba, S.U. (2015). A study on the individual impact of cassava starch and hydroxyl propyl-modified starch on mud density. International Journal of Engineering*) and (B) schematic of the density balance.

The mixer motor is then turned on and maintained at a constant speed of 4000 ± 250 RPM, and the dry cement is slowly added to the mixing container at a uniform rate for no more than 15 seconds. After 15 seconds at 4000 ± 250 RPM, the speed of the mixer is increased to 12,000 ± 500 RPM, and mixing is continued to another 35 ± 1 seconds. The slurry prepared is then tested for properties such as density, rheology, and thickening time, or poured into molds for setting and aging before being tested for other properties.

3.3 Slurry density

To ensure the proper slurry characteristics in the field and downhole, the density of the cement slurry must be measured under controlled laboratory conditions. Measuring density of the cement slurry is typically performed using a pressurized fluid density balance. Based on API 10B specifications the apparatus consists of two major components—a thin syringe-like pump and the density balance—both of which are shown in Fig. 3.2. The tool utilizes a syringe-like mechanism to pressurize the slurry in the cup,

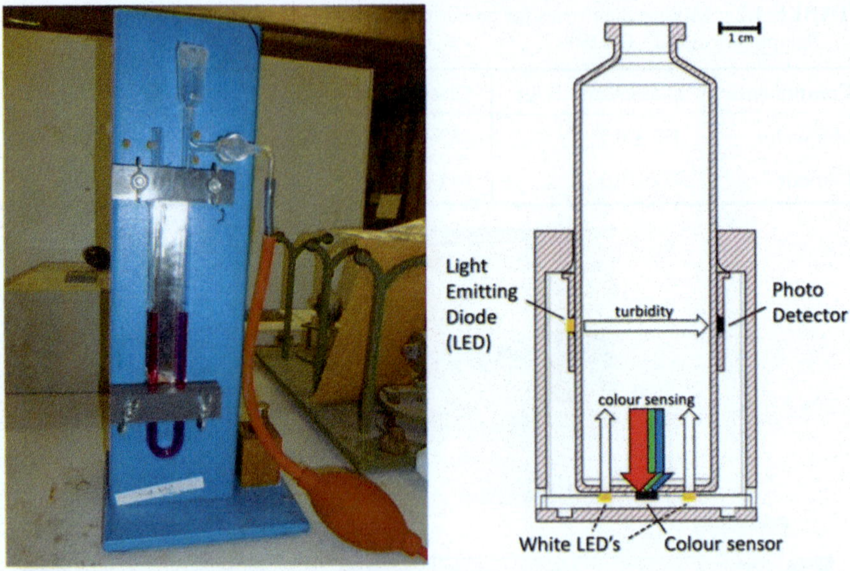

FIGURE 3.3 (A) Blaine permeameter, From Michel and Courard (2014) and (B) Schematic of a turbidimeter to measure fineness, From Barbe et al. (2022).

effectively removing any entrained air, before it is placed on the density balance.

Once loaded on the density balance, the counterweight is balanced using the slide and scale to level the tool, and the readings at the point of balance are recorded (Devers, 2024). Density control is typically performed by altering the water content in the cement or by the addition of extenders such as bentonite. The density of cement slurries used in oilfield applications can vary from 8.33 to even 20 ppg for high-pressure formations, and the slurry density used depends on the specific downhole conditions, from normal pressure and temperature to controlling abnormal formation pressures (Drilling Manual, 2021).

3.4 Fineness test

Fineness test is typically conducted on the neat cement samples to control cement quality used in oil and gas applications. This is a standardized test based on the API 10A, ASTM C115, and ASTM C204 standards. The fineness test determines the surface area of cement particles and can be measured using either a turbidimeter or an air permeameter and is expressed in meters squared per gram (m^2/g).

For measuring cement fineness using a permeameter (Fig. 3.3A) as per the ASTM C204 standards, neat cement power is compressed into

TABLE 3.3 Requirements of fineness for different cement classes, Adapted from API (2019).

Fineness tests						
Well cement class	**A**	**B**	**C**	**D**	**G**	**H**
Turbidimeter (specific surface, minimum, m^2/kg)	150	160	220	NR	NR	NR
Air permeability (specific surface, minimum, m^2/kg)	280	280	400	NR	NR	NR

NR, *no requirement.*

a porous plug, and air or neutral gases are passed through the cement plug sample at specific rates. The surface area of the particles can then be determined by calculating the resistance to airflow (ASTM, 2024). When measuring fineness using a turbidimeter (Fig. 3.3B), ASTM C115 standards are applied, and a turbidimeter is used to calculate the specific surface area of cement. A source of light of specific constant intensity is passed through a suspension of cement onto a photoelectric cell, and the current generated is measured using a microammeter to determine the turbidity of the suspension (ASTM, 2010).

Table 3.3 shows the API requirements of (acceptable) fineness for the turbidimeter as well as for the permeameter. Classes D, G, and H cements have no fineness requirements. Finer cement also allows for faster hydration, leading to greater strength and a shorter cement setting time. Finer cement is also preferred because it has lower porosity and permeability and thereby ensures better well integrity and is better suited for high-pressure high-temperature (HPHT) situations.

3.5 Consistency or thickening time

Lab testing is performed to ensure that the cement slurry remains fluid during mixing as well as pumping into the well. The cement slurry must remain mobile during its flow through the tubing/casing and displacement through the annulus. A pressurized consistometer is utilized for testing the mud sample, which is placed in an HPHT vessel that has a diaphragm and a paddle that can rotate freely. A heating system capable of raising and controlling the temperature of the setup is also required, whereas the slurry container is rotated to measure consistency. API (2019) provides further specifications for the consistometer required and the detailed steps required in taking consistency measurements. Fig. 3.4 shows images of an industrial consistometer.

Consistency measurements of the cement slurry being tested are expressed in Bearden units (B_c), which are dimensionless. During the test

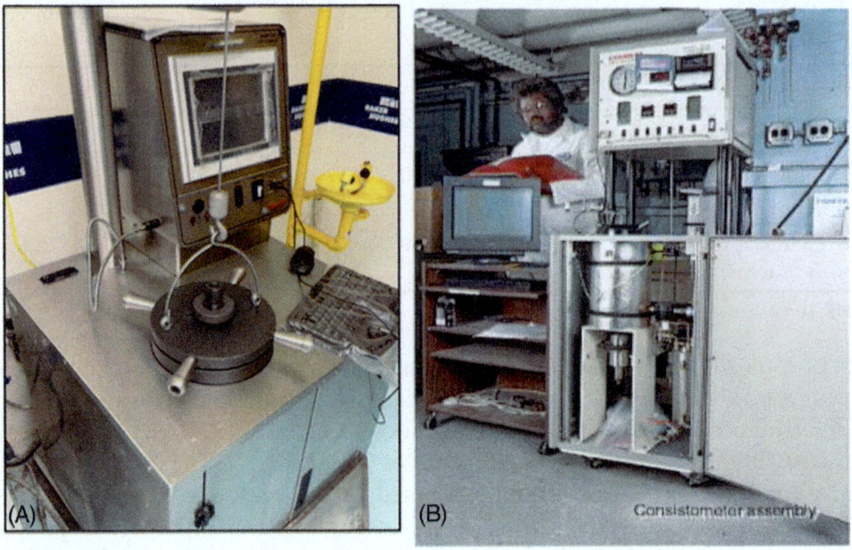

FIGURE 3.4 (A) Consistometer setup, Adapted from Helmy and Fakher (2024) and (B) sample consistometer assembly for HPHT measurements (Wagh, 2016).

period, the temperature and pressure of the cement slurry are increased consistently until the consistometer reading shows a consistency of 100 B_c, and the time required to reach this consistency is recorded as the thickening time for the slurry. The time during which the consistency of the slurry remains below 30–40 B_c is generally considered the pumpable time, and by the time the slurry has a consistency of over 70 B_c, the slurry has thickened enough to a point where it will be difficult to move in a borehole. Fig. 3.5 shows a sample trace of the consistency of a cement slurry with temperature over time.

As per API 10A the recommended thickening time standards required for neat samples of different cement classes are shown in Table 3.4. These are the minimum and maximum recommended thickening times as per API, though different companies can have different specifications or requirements as per the job. In fact, cement thickening times in the field can vary from 20 to 30 minutes to days depending on the pressure and temperature downhole, cement class, formulation or additives used, and the mixing conditions of the cement slurry.

3.6 Rheology test—Viscosity and gel strength

Measuring and understanding the rheology of the cement slurry are necessary to evaluate its efficiency for pumping and to ensure proper placement in the wellbore. Viscosity is very important in cementing operations,

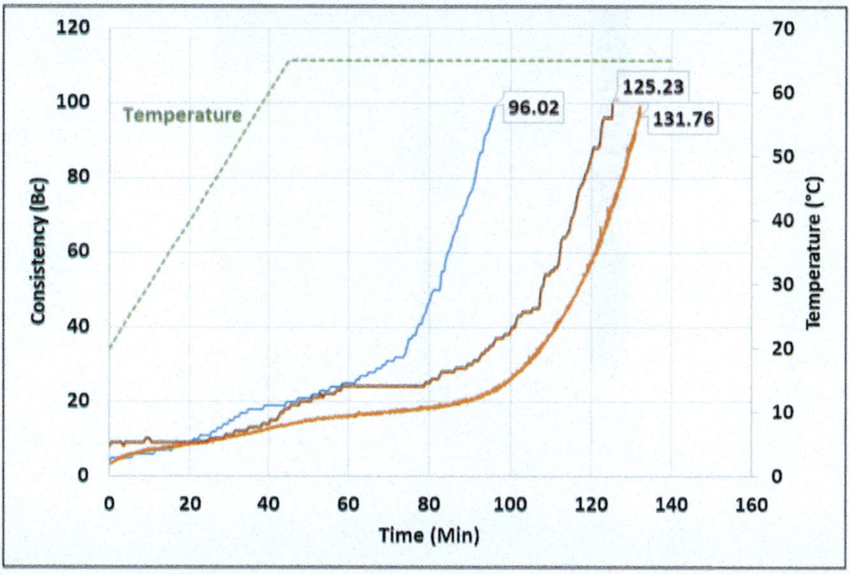

FIGURE 3.5 Consistometer readings of the thickening time for a sample cement slurry, Adapted from Kamali et al. (2021).

TABLE 3.4 Requirements of thickening time for different neat cement classes (API, 2019).

Cement class	Schedule	Minimum thickening time (min)	Maximum thickening time (min)
A	4	90	NR
B	4	90	NR
C	4	90	NR
D	4	90	NR
D	6	100	NR
G	5	90	120
H	5	90	120

NR, *no requirement*.

as it determines the pumping efficiency of the cement, as well as fluid displacement in the wellbore. Low-viscosity cement will require higher flow rates to have better displacement, whereas high-viscosity cement will display better displacement at lower flow rates (Drilling Manual, 2021).

API Specification 10B has developed a standardized procedure to determine the rheology of the cement slurry used for oil and gas applications

FIGURE 3.6 Fann rotational viscometer, used to measure viscosity, Adapted from Yan et al. (2021).

because these properties are sensitive to the measurement procedure followed. Fig. 3.6 shows a rotational viscometer, which is used for measuring rheological properties and consists of two concentric cylinders rotating at a constant RPM, with the slurry sample confined between them. Information obtained from the rotational viscometer can be used to determine slurry viscosity, gel strength, shear rate, shear stress, and the yield point.

The viscometer can measure the rheological properties at ambient pressures and temperatures, as well as under downhole conditions, including HPHT scenarios. Cement slurry to be tested is poured into the viscometer slurry cup to the fill line, and the test should be initiated immediately at the lowest RPM setting to prevent the slurry from remaining static. In the case that the cement slurry must be tested at higher temperatures or pressures, a consistometer is used to condition the cement slurry to the desired temperature and pressure before being loaded into the viscometer slurry cup. Viscometer readings should be taken at RPM settings of 3, 6, 30, 60, 100, 200, and 300 RPM, keeping the temperatures constant, as shown in example Table 3.5 (API, 2013). Further rheological properties and flow behavior can be characterized from this data using different flow rheological models (Bingham Plastic, Power Law, Herschel–Bulkley, etc.). Note that cement viscosity can be adjusted by adding water to the mixture, as well as incorporating dispersants.

TABLE 3.5 A sample test results for the viscometer test, Adapted from API (2013).

RPM	Ramp-up readings	Ramp-down readings	Reading ratio	Average readings
3	21	24	0.87	22.5
6	40	36	1.11	38
30	65	83	0.78	74
60	84	100	0.84	92
100	100	115	0.87	107.5
200	137	147	0.93	142
300	170			170

Initial slurry temperature = 150 F; final slurry temperature = 146 F; rheological properties reported at average temperature = 148 F.

Gel strength is determined immediately following the viscometer test for rheology at 300 RPM. The 10-second and 10-minute gel strengths are both determined as per API Specification 10B. The viscometer is shut off for 10 seconds before rotation is restarted at 3 RPM, and the maximum observed reading is noted immediately after. This information is used to calculate the 10-second gel strength. The viscometer is then shut off for 10 minutes, after which it is restarted at 3 RPM, and the maximum observed reading is noted to determine the 10-minute gel strength. The gel strengths should be reported at the temperature at which they were tested (API, 2013).

3.7 Free-water content test

The *free-fluid* or *free-water* test is conducted to assess the level of separation between the water and the mixed solids in the cement slurry, including cement and additives. This test indicates the stability of the cement formulation. A cement slurry with a high free fluid content is not recommended for operations. All cement slurries, even after proper mixing, will have some amount of free fluid, which can cause separation and channeling in the cement layer, leading to migration of oil or gas through these channels and finally a loss in containment (Drilling Manual, 2021). The free fluid in the cement slurry is measured following the guidelines outlined in API Specification 10A.

Before starting the test, cement slurry is stirred in the consistometer to homogenize for a period of 20 minutes. A known volume of the slurry is then transferred to a conical flask, which is then sealed to prevent evaporation. The slurry-filled flask is then left undisturbed for a period of 2 hours ± 5 minutes. At the end of the resting period, the free fluid that has developed (i.e., separated from the slurry) is removed with a syringe or a pipette, and its volume is measured. This data can then be used to calculate the volume fraction of free fluid in the cement slurry, as explained in API Specifications 10A. The API requirements specify that the free fluid for Classes G and H oil well cements should not exceed 5.9% of the slurry volume (API, 2019).

In highly deviated or horizontal wellbores the free-water level in the cement slurry becomes more critical. In these types of wells, cement stability is even more important for effective zonal isolation. Free water can negatively affect the slurry's consistency, leading to poor bonding, uneven placement, or even potentially lost circulation. Therefore managing the free-water content would ensure the slurry stability throughout the placement process and is especially important in these challenging wellbore geometries.

3.8 Fluid loss test

Cement slurries must be tested for fluid separation and loss into formation pores under various conditions before being pumped downhole. Excessive loss of slurry fluid can cause a loss of cement hydration, leading to subpar cementing quality and diminished ultimate strength of the cement. Fluid loss tests are conducted generally at the temperature exposed bottomhole (bottomhole circulating temperature), and the cement samples are gradually prepared for these conditions before being tested. The API Specification 10B should be followed for fluid loss testing, and there are guidelines for performing the tests at temperatures less than 90°C and above 90°C. An API fluid loss setup, or an HPHT fluid loss setup is used for the test, and the fluid loss values are determined in static conditions. The fluid loss cell is capable of being pressurized and is fitted with a 325-mesh screen, with a 60-mesh size backup screen (Fig. 3.7).

Before testing the slurry should be mixed according to API standards and with the additives required for field use. Following this the conditioned slurry is placed in the fluid loss cell, and the cell is pressurized. A pressure differential of 500 ± 50 psig is applied across the cell, and the slurry is agitated and heated gradually to the bottomhole circulating temperature and according to the thickening time that is noted in the field. Once proper agitation and heating have been achieved, increase the pressure differential to 1000 ± 50 psig, and then open the bottom fluid collection valve. The filtrate is collected, and the volume is continuously recorded over varying time periods, ranging from 30 seconds to 30 minutes. The API fluid loss is then determined by using the collected filtrate

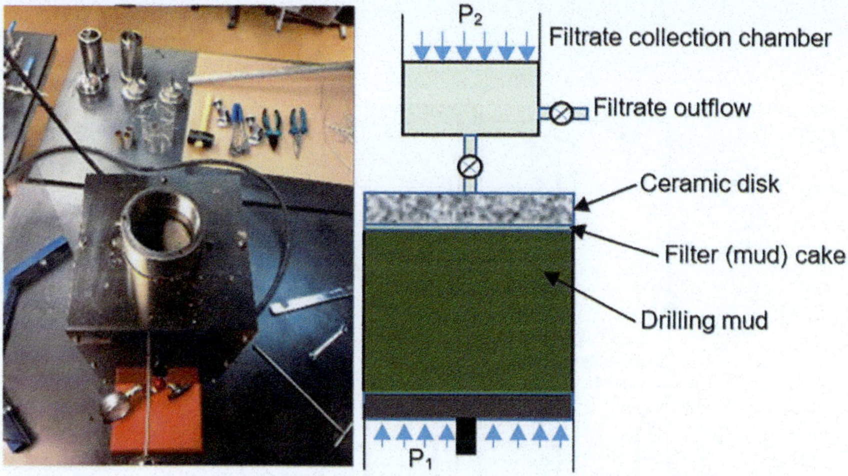

FIGURE 3.7 API fluid loss measurement equipment for normal and HPHT conditions and the schematic of a filtrate loss setup, Adapted from Kanjirakat et al. (2026).

volume at the end of 30 minutes and doubling that value (API, 2013). Generally, for stable cement slurries used in oil and gas wells, the cumulative fluid loss should be less than 100 mL for standard operations and less than 50 mL for wells with gas influx (where a more stringent limit is needed). Note that fluid loss control may be achieved using additives such as bentonite, cellulose, or high molecular weight polymers in the cement slurry (Drilling Manual, 2022).

3.9 Permeability test

Cements used in offshore deepwater wells and gas-producing wells face increased risks of seepage and migration, which can potentially travel up the cement layer to the surface. The API Specification 10B outlines recommended procedures for measuring the permeability of cement sheath. This ensures that the risk of gas or oil migration through the cement layer is minimized. The results of these tests are crucial for refining cement formulations and improving well design, helping to prevent potential issues in the future, but the actual permeability of the set cement in downhole conditions cannot be completely and reliably determined through laboratory testing.

Permeability measurements are carried out using a cement or a core permeameter, as shown in Fig. 3.8, in which a cement sample is loaded, and a gas or liquid is passed through the sample while measuring the pressure and flow rate through the sample. The cement sample can be a molded

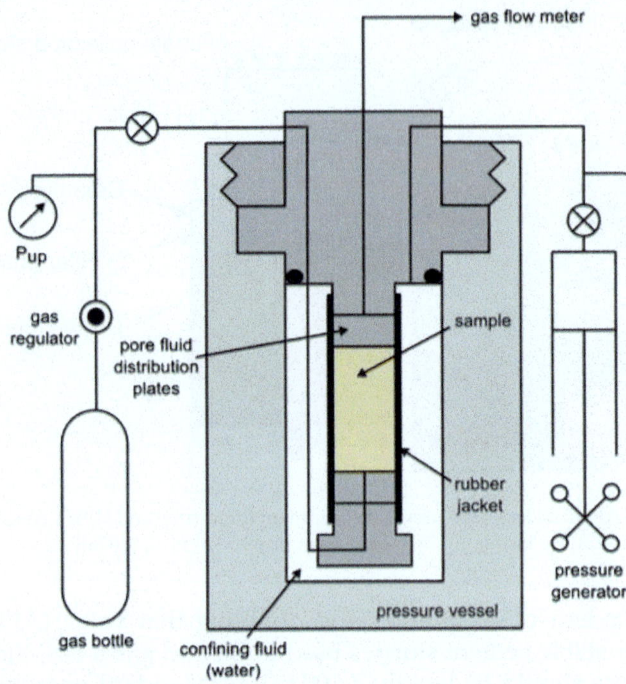

FIGURE 3.8 Schematic of a permeameter, Courtesy Siratovich et al. (2014).

sample or a cored sample, and there are different types of permeameters available for each type. To measure liquid permeability the cement sample is initially saturated with water, and an accumulator is used to slowly flow water through the sample. The inlet pressure and flow rate are measured, and then Darcy's law is applied to calculate liquid permeability, as shown in Eq. (3.1) (API, 2013).

$$K = 14,700\,\frac{Q\mu L}{A\Delta P} \tag{3.1}$$

where Q is gas flow rate (mL/second); μ is viscosity (cp); L is length (cm); A is cross-sectional area (cm^2); ΔP is the pressure differential (psi); and K is permeability (mD).

To measure gas permeability the sample should be dried in an oven to a constant weight before being loaded to a permeameter. A confining pressure is applied to the sample to simulate hydrostatic pressure seen bottomhole, and a pressure differential can be applied using a backpressure regulator. Gas, most commonly nitrogen or helium, is passed through the sample, and the pressures on the inlet and outlet gauges are noted. The flow rate is measured using a flow meter once it has stabilized, and this data is used to calculate the permeability using Darcy's law, as per Eq. (3.2). The API Specifications 10B provide additional comprehensive

FIGURE 3.9 Compressive strength testing using stress testing equipment, Adapted from Abraham et al. (2024).

information on sample preparation, loading, and test procedures related to permeability measurements (API, 2013).

$$K = 2000\,\frac{Q_b \mu P_b L}{A\left(P_i^2 - P_o^2\right)} \tag{3.2}$$

where Q_b is flow rate (mL/second); μ is viscosity (cp); L is length (cm); P_b is adjusted pressure (atm); A is cross-sectional area (cm^2); P_i is inlet pressure (atm); P_o is outlet pressure (atm); and K is permeability (mD).

3.10 Destructive compressive strength test

Cement slurry develops strength over time as it slowly sets, and this strength is generally a function of the temperature, curing time, and pressure. Measuring the compressive strength of the cement sample is of utmost importance to determine its failure criteria, and this is commonly performed by a uniaxial compressive stress test. This stress test is performed until the sample deforms and fails permanently when subjected to mechanical pressure. The procedure for performing this test is guided by the API 10A and the ASTM C109 guidelines. The cement samples must be prepared in cubical molds with size 50 mm (2 in) with tight error tolerances. The samples can then be cured in atmospheric or pressurized water baths, which can be at room temperature or higher temperatures. The samples must be cured for the minimum specified curing periods before they can be tested, as listed in API 10A specifications (API, 2019).

To conduct destructive compressive testing on cement samples the samples are positioned in the testing equipment, centered beneath the block that applies the compressive stress, as illustrated in Fig. 3.9. Stress is then applied to the sample at a constant loading rate of about 18 kN/minute

TABLE 3.6 Minimum compressive strength requirement as prescribed by API 10A (API, 2019).

			Minimum compressive strength at indicated curing period	
Cement class	**Final curing temperature**	**Curing pressure MPa (lbf/in.^{2})**	**8 h ± 15 min MPa (lbf/in.^{2})**	**24 h ± 15 min MPa (lbf/in.^{2})**
A	38 (100)	atmospheric	1.7 (250)	12.4 (1800)
B	38 (100)	atmospheric	1.4 (200)	10.3 (1500)
C	38 (100)	atmospheric	2.1 (300)	13.8 (2000)
D	77 (170)	20.7 (3000)	NR	6.9 (1000)
	110 (230)	20.7 (3000)	3.4 (500)	13.8 (2000)
G, H	38 (100)	Atmospheric	2.1 (300)	NR
	60 (140)	Atmospheric	10.3 (1500)	NR

NR, *no requirement.*

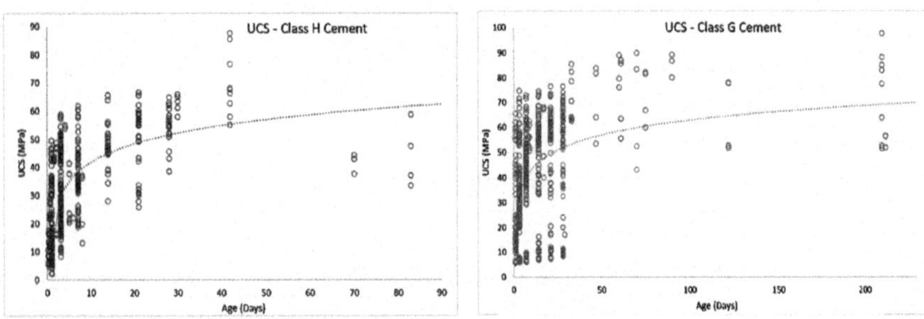

FIGURE 3.10 Compressive strength of Class H and Class G cements over time, Adapted from Abraham et al. (2024).

(4000 lbf/minute) for samples aged for 8 hours and about 72 kN/minute (16,000 lbf/minute) for all other samples as per the API specifications. The uniaxial compressive strength (UCS) is found as the minimum force required to break the sample mechanically and is usually expressed in megapascals (MPa) or in pound-force per square inch (lbf/in^2). Table 3.6 shows the requirements for minimum compressive strength as specified by API 10A.

Neat, clean cement will generally have the greatest compressive strength, whereas the addition of loss control agents, retarders, accelerators, and other additives can generally impair the ultimate compressive strength of cement. Fig. 3.10 shows an example of the progression of compressive strength of cement over time. Higher temperatures during

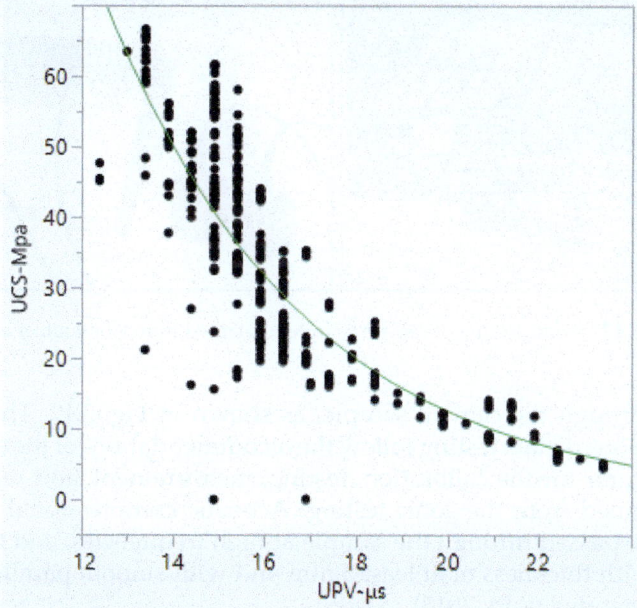

FIGURE 3.11 Relationship between compressive strength and sonic transit time for a Class H cement sample, Adapted from Abraham et al. (2023).

curing are also noted to impair the compressive strength of cement—a phenomenon commonly referred to as cement strength retrogression, and in HPHT conditions, additives, such as silica, are required to prevent this impairment (Abraham et al., 2024).

3.11 Nondestructive sonic testing

The compressive strength of cement can also be evaluated using non-destructive methods, such as ultrasonic testing (ultrasonic pulse velocity). The sonic compressional wave transit time has a strong relationship with the ultimate compressive strength of the sample, as shown in Fig. 3.11, and can be used to get an accurate measurement of the compressive strength without damaging or deforming the sample. The API Specifications 10B generally govern the standards for performing this test, but note that the UCS measured using sonic velocities cannot be accurately reported as actual compressive strength, as it can be influenced by the free fluid level, sample geometry, and changes in temperature and pressure (API, 2013).

The apparatus required for sonic testing consists of equipment that can emit acoustic signals and transducers to transmit and measure the

FIGURE 3.12 Sonic testing of cement samples, Adapted from Abraham et al. (2023).

signals through the cement sample, as shown in Fig. 3.12. The API 10B specifications dictate testing follow the recommendations of the equipment manufacturer for the calibration, testing, measurement, and recording of data obtained from the sonic testing. Acoustic compressional and shear waves are passed through the sample at high frequencies, and therefore a sample with thickness of at least 5 mm and with smooth parallel surfaces is recommended (API, 2013).

3.12 Thermal conductivity

Finding the thermal conductivity of different cement formulations is crucial for geothermal applications as well as high-pressure, high-temperature (HPHT) operations. To ensure optimal performance in such demanding environments, properly maintaining a thermally insulated environment is necessary across nonreservoir intervals (most of the wellbore) so that the superheated steam can return to the surface with little loss of thermal energy to ensure maximum efficiency. Conversely, across geothermal or HPHT reservoir zones, a highly conductive cement layer (thermally) is necessary to ensure that the heat transfer is highly efficient to superheat the steam the most (Abid et al., 2023).

Cements typically have low thermal conductivity, ranging from 0.18 to 1 W/m/K (2.6–13.8 Btu/ft/day/°F). To overcome the thermal challenges associated with using traditional cements across reservoir zones, it is essential to enhance the thermal conductivity of cement blends. Although this concept has not been widely explored in the drilling industry, improving thermal conductivity could prove to be a crucial solution (Ashena et al., 2023).

Measuring thermal conductivity is generally performed using a test platform such as the Thermtest Portable Measurement Platform (MP2), which consists of a probe and a conductivity meter such as a Keithly 2400,

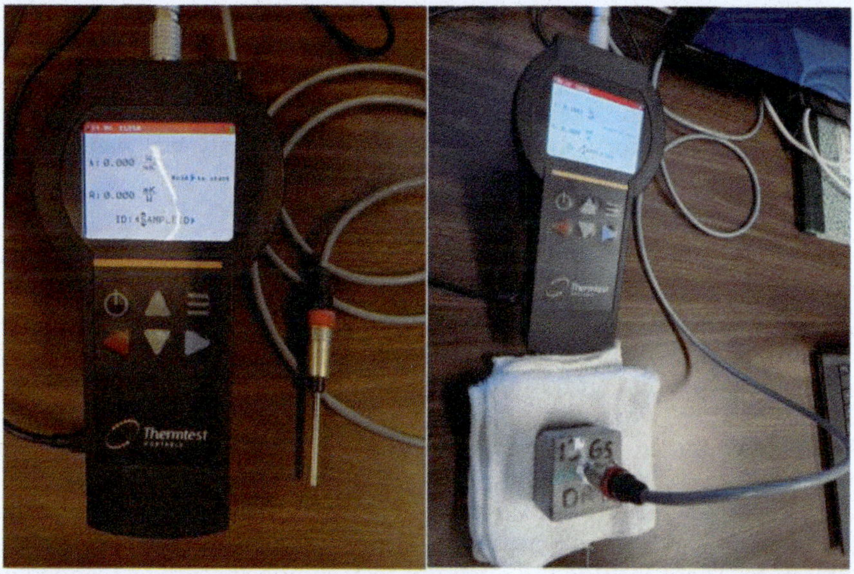

FIGURE 3.13 Thermal conductivity measurement, Adapted from Abid et al. (2023).

as shown in Fig. 3.13. A hole—typically 4 mm in diameter—is drilled into the cement sample, and the probe is placed in the hole. Heat signals are sent through the sample, and then the sensor measures the reading on the outer surface of the cement sample.

3.13 Cement bond strength tests

Given the critical role of the cement sheath in ensuring well integrity, it is essential to evaluate and quantify the bond integrity between the cement and both the formation and the casing. Though there are no standardized tests specified by API or other institutions, work has been done by several researchers to devise ways of determining the bond strength; therefore the shear bond test, tensile bond test, and hydraulic bond test are considered the most critical methods for evaluating downhole cement. Among these the shear bond strength (SBS) test is the most commonly used and widely recognized laboratory method for assessing bond strength. This section will explain these methods.

Although these tests provide reliable results in a laboratory setting, their accuracy in the downhole environment is not fully guaranteed. The lack of unified standards for evaluating the cement bonding interface can lead to discrepancies and potential errors in assessing the bond's performance under actual downhole conditions, and there are also significant limitations to simulating the real downhole bonding interface in the laboratory. The complexities of the downhole environment in the presence of temperature, pressure, well fluids, as well as other wellbore and formation conditions, provide less than ideal environment to ensure proper cement bonding; as a result, the laboratory-obtained results may not be fully applicable (Dong et al., 2024).

3.13.1 Shear bond test

SBS is defined as the mechanical strength needed to support the pipe in the hole. It is determined by assessing the force required to break the cement-casing bond and displace the casing string (Carter & Evans 1964). In more general words, the SBS is generally defined as the minimum force required to initiate a relative displacement between the cement layer at either the casing or the formation. As there is no universally recognized standard methodology for conducting a shear bond test, such as those outlined by API or NORSOK, several recommended methods exist in the literature. These include the push-out test, composite core test, and slant shear test (Dong et al., 2024). These tests are typically performed on core-sized cement samples that interface with both the casing and the formation rock.

3.13.1.1 Push-out method

Of the various methods the push-out test has been the most commonly used for several decades to determine the SBS of common cement formulations and the casing or the formation rock layer (Evans & Carter, 1962; Ladva et al., 2005). Depending on the interface to be tested, in this test, either a cylindrical steel core or a cylindrical formation rock sample is placed inside a slurry mold, and a plastic spacer is placed on top to isolate the rock or steel core, as shown in Fig. 3.14. The cement slurry, formulated as per API 10A/10B specifications, is poured into the mold around the rock or steel core and is allowed to set and cure under different conditions like those experienced downhole for a period of at least 7 days. This setup helps simulate the bonding interface between the cement and the casing or formation, similar to what occurs in downhole conditions.

Following the curing period the mold is removed, and the cement with the steel or rock core is placed upside down on an axial loading stand. The stand is designed to only hold on to the cement layer, whereas the inner core is free to move when load is applied. A hydraulic pressure, similar to those used for axial loading, is then applied as an increasing vertical load on the steel or rock core by pushing on the core (using a spacer or a rod) until the core debonds and is displaced from the cement layer. The failure load is noted, and the bond strength is then calculated from the force

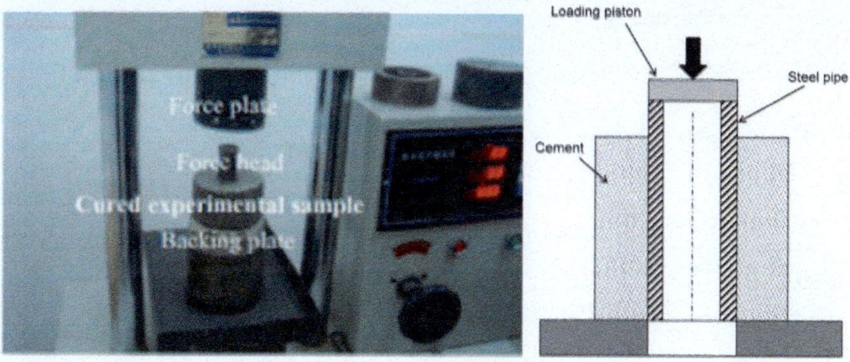

FIGURE 3.14 Push-out test to determine bond strength, Adapted from (Left) Dong et al. (2024); and (Right) Lavrov et al. (2019).

required for displacement using Eq. (3.3):

$$P = \frac{F}{S} = \frac{F}{2\pi rh} \tag{3.3}$$

where P is the bond strength between the cement and casing interface (MPa), F is the failure load (N); S is the contact area (m^2); and h is the height (m). Although the push-out method has several advantages, including being simple, easy to design and testable at different pressures, and temperatures. However, it also comes with several limitations, including poor reproducibility, shear failure not happening at the cement–rock/cement–steel interface, or chances for tensile failure or thermal contraction—all of which will skew the SBS calculated from the test (Liu et al., 2015).

3.13.1.2 *Composite-core method*

The composite-core method was proposed by Agbasimalo and Radonjic (2014) as a different approach to finding SBS. In this method, a composite core with cement on one side and formation rock on the other side for a total length of 2 in is used, as shown in Fig. 3.15. The interface between the two surfaces mimics bonding between the materials as seen downhole. A load is applied on the composite core in a direction parallel to the interface, and failure is expected to occur at the bonding interface. This failure stress is noted, using which the bond strength can be calculated. Although tests conducted using this method are simple and have shown to give reproducible results, this technique was found unsuitable for rocks with laminated layers like clay/shale because in this case the failure would happen in those layers instead of the composite-core interface (Liu et al., 2015).

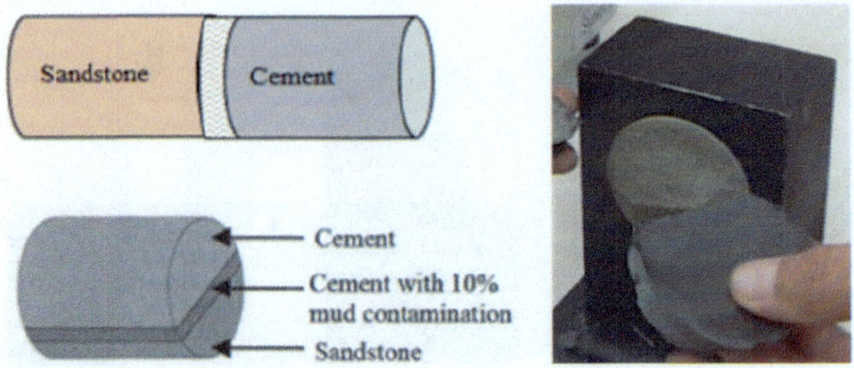

FIGURE 3.15 Composite-core test to determine shear bond strength after interface failure in a composite-core, Adapted from Dong et al. (2024).

3.13.1.3 *Slant shear test*

The slant shear test, initially proposed by Liu et al. (2018), utilizes a common test method to measure SBS of concrete, as is the case in civil and construction industry. For this test, first, a core sample of the rock is cut into two halves at a 60-degree angle. One half of the core is then placed in a mold, and the cement slurry is poured on the top of the core sample and allowed to set to create a composite core consisting of both the rock and the cement layer, as shown in Fig. 3.16. The composite core is then cut from either end to ensure that the length-to-diameter ratio is 2:1, while taking care to not damage the bonding interface between the cement and the rock. A mechanical load is then applied on the composite core vertically until it deforms and fails, and this load is used to calculate the bonding strength (Dong et al., 2024):

$$\sigma_s = \frac{4F_{\text{peak}} \sin 60 \cos 60}{\pi d^2} \tag{3.4}$$

where σ_s is the bonding strength at the interface (MPa), F_{peak} is the load applied at the interface on failure (N); A is the contact area (m^2); and d is the core diameter (m). The interface angle between the cement and the rock core is kept at 60 degrees to identify if failure is occurring at the interface or in the rock or cement, and to ensure that only a failure occurring at the cement–rock bonding interface is accounted for in calculating the bond strength.

3.13.2 Hydraulic bond test

Although the shear bond test is most commonly used to determine the bond strength at the interface, it does not take into account the presence

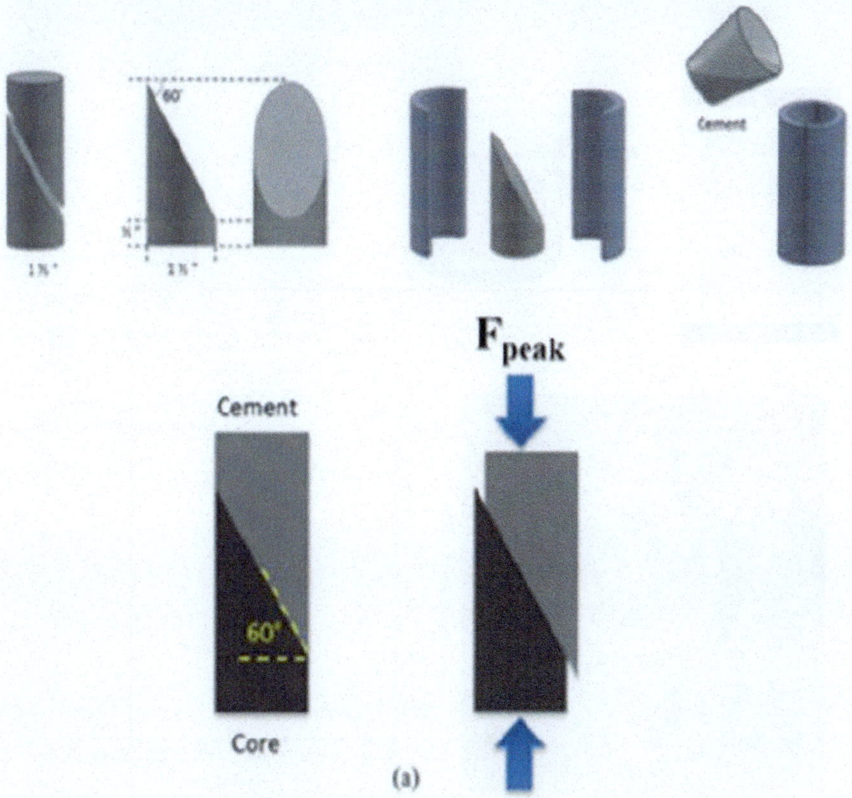

FIGURE 3.16 Slant shear test to determine the bonding strength, Adapted from Dong et al. (2024).

of fluids, which can contribute to a radial debonding around the cement-casing/formation. This is the most common form of a seal failure downhole and is accelerated when the fluids are exposed to a range of temperatures and pressures—as experienced during a drilling and production operation. So, the hydraulic bond test involves applying pressure to simulate downhole conditions and measuring the ability of the cement bond to resist fluid migration. The test ensures that the cement bond is strong enough to block the flow of formation fluids around the cemented zone, which is crucial for maintaining well integrity and preventing issues like fluid leaks. Thus determining the hydraulic bond strength is important to ensure genuine cement sheath integrity (Yang et al., 2020). The hydraulic bond test is conducted to assess the strength of the cement-casing/formation bond at the interface, ensuring it is strong enough to prevent migration of formation fluids around the cemented layer. A hydraulic bond test is similar to the push-out shear bond test but involves injecting a fluid around

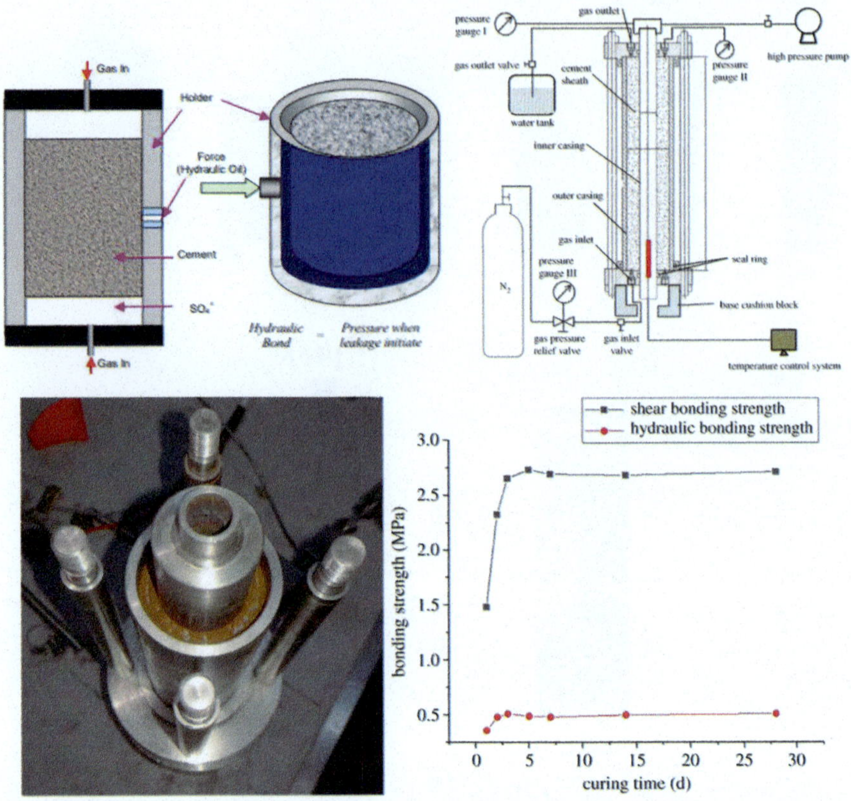

FIGURE 3.17 Experimental setup of a hydraulic bond test apparatus and the hydraulic bond test pressure profile; (Top-left) From Condor and Asghari (2008); (Top-right, Bottom-left, and Bottom-right) From Yang et al. (2020).

the cement-casing/formation interface at pressure to initiate debonding and continues until leakage occurs. Again, this test is aimed at determining the minimum required pressure to initiate flow along the interface between the two surfaces—indicating failure. Though there is no standard method to perform this test, it usually involves a cement plug set inside a casing string, and fluid is injected at pressure until flow is initiated at the interface. Fig. 3.17 shows a schematic of a typical setup used for hydraulic bond testing (Condor & Asghari, 2008; Yang et al., 2020).

For the hydraulic bond test the setup consists of a high-pressure steel container, which has inlets for the connection of pumps and injection of fluids, pressure gauges, a gas or liquid flowmeter, and valves to control the flow. Note that hydraulic bond tests can be conducted with liquids or gases, depending on the type of formation fluid downhole, and the equipment can be adapted to have a pressurized chamber for gas flow if required, as

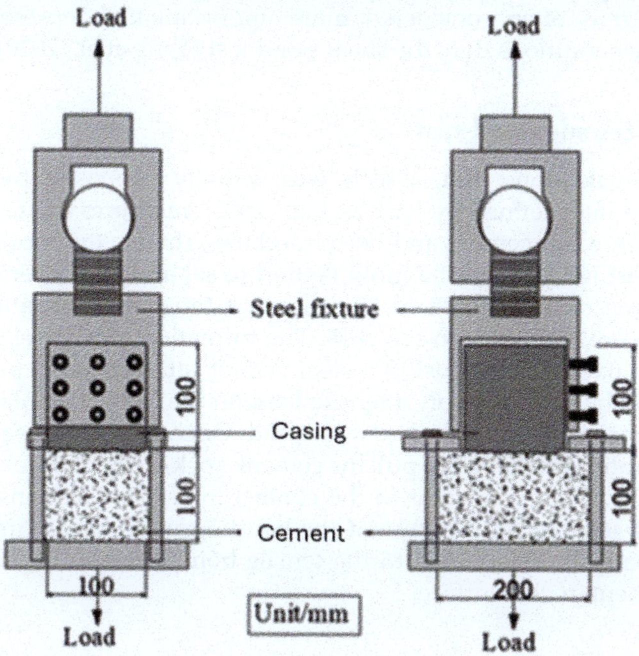

FIGURE 3.18 Tensile bond test setup, Modified from Wang et al. (2023).

shown in Fig. 3.17. Cement slurry is prepared as per API specifications and according to the requirements of the field and then poured into the steel casing/container to be tested. The inlet and outlet valves for fluid flow are kept closed during the cement pouring and curing phases, and the cement can be cured at pressures or temperatures typically seen downhole using an oven. Once the cement is cured, the valves are opened and liquid/gas is gradually injected into the container—at the cement–steel interface. Injection pressures are monitored continuously until flow occurs, and the flowmeter starts recording flow rate values. The test is continued until the liquid/gas flow readings become stable, and this pressure is noted as the hydraulic bond strength (Gomado et al., 2022; Yang et al., 2020).

As no universal standard currently exists for hydraulic bond testing, the setup design is open to some arbitrary selection, with the primary goal being the injection of fluids at the cement-casing/formation interface. The fluids used in the test can include deionized water, saline water, oil, or other suitable liquids, as well as gases such as nitrogen, air, or helium. The injection of fluids is controlled by the pumps, and the pressure is slowly increased in 5–10-minute intervals until flow is stabilized (Khalifeh et al., 2018). As the hydraulic bond test takes into account the presence of formation fluids and their role in cement-casing/formation integrity, it is often considered more important and representative of downhole conditions than the shear bond test (Yang et al., 2020).

3.13.3 Tensile bond test

The tensile bond test, which was initially proposed by Peterson (1963) and later refined by Ladva et al. (2005), measures the tensile bond strength between cement and both interfaces, that is, the formation rock or steel casing. In this test a mold is used to set and cure a cement slurry prepared according to API specifications on top of the formation rock or the steel core, as shown in Fig. 3.18. The formation rock/steel core at the bottom is fixed in place using a steel anchor attachment. A pulling rod, which will be used to apply a tensile load on the cement, is placed partly inside the cement slurry before it is cured. Once completely set, a tensile load is applied on the rod to pull the cement–rock/cement–steel composite in a direction perpendicular to the contact interface. This tensile load is increased steadily until failure at the bonding interface occurs, and this failure load can be equated to the tensile bond strength of the cement–steel/cement–rock interface.

References

Abid, K., Srivastava, S., Tellez, M., Amani, M., & Teodoriu, C. (2023). Experimental and machine learning study of thermal conductivity of cement composites for geothermal wells. *Geothermics, 110*, 102659. https://doi.org/10.1016/j.geothermics.2023.102659.

Abraham, J., Devers, C., Teodoriu, C., & Amani, M. (2023). A comprehensive evaluation of properties of common and regional cement formulations from a well integrity standpoint. In *Offshore technology conference. OTC-32419-MS*. Houston, Texas: Society of Petroleum Engineers https://dx.doi.org/10.4043/32419-MS.

Abraham, J., Devers, C., Teodoriu, C., & Amani, M. (2024). Stress failure assessment and potential well integrity issues with different oilwell cement classes and formulations in complex wells. In *2024 Gas and Oil Technology Showcase and Conference (GOTECH), Dubai, UAE* https://doi.org/10.2118/219117-MS.

Abraham, J. J., Devers, C., Teodoriu, C., Amani, M., & AlMujalhem, M. (2025). Evaluating thermal stability of geopolymers and other common cement formulations in geothermal and other HPHT operations. In *Paper presented at the Offshore Technology Conference, Houston, Texas, USA* May 2025 https://doi-org.ezproxy.lib.ou.edu/10.4043/35841-MS .

Agbasimalo, N., & Radonjic, M. (2014). Experimental study of the impact of drilling fluid contamination on the integrity of cement–formation interface. *Journal of Energy Resources Technology, 136*(4), 042908. https://doi.org/10.1115/1.4027566.

Ashena, R., Madani, M., Sivanesan, S. K., & Vinesh, T. (2023). Investigating the effect of thermal conductivity on geothermal energy production at different circulation rates in an EGS abandoned case study. In *Paper presented at the International Petroleum Technology Conference, Bangkok, Thailand* March 2023 https://doi.org/10.2523/IPTC-22762-EA .

API. (2013). *Recommended practice for testing well cements – API Specifications 10B*. American Petroleum Institute.

API. (2019). *Cements and materials for well cementing – API Specification 10A*. American Petroleum Institute.

API. (2024). *API publications catalogue 2024*. API Retrieved from https://www.api.org/-/media/Files/Publications/2024-Catalog/2024-Publication-Catalog.pdf .

ASTM. (2010). *Standard test method for fineness of Portland cement by the turbidimeter*. American Society of Testing and Materials.

ASTM. (2024). *Standard test methods for fineness of hydraulic cement by air-permeability apparatus*. American Society of Testing and Materials.

Barbé, B., Corsmit, E., Jans, J., Kaur, K., Baets, R., Jacobs, J., & Hardy, L. (2022). Pilot testing of the "turbidimeter", a simple, universal reader intended to complement and enhance bacterial growth detection in manual blood culture systems in low-resource settings. *Diagnostics, 12*(3), 615. https://doi.org/10.3390/diagnostics12030615.

Carter, L., & Evans, G. (1964). A study of cement-pipe bonding. *Journal of Petroleum Technology, 16*(02), 157–160. https://doi.org/10.2118/764-PA.

Condor, J., & Asghari, K. (2008). Experimental study of stability and integrity of cement in wellbores used for CO_2 storage. *Energy Procedia, 1*, 3633–3640. https://doi.org/10.1016/j.egypro.2009.02.159.

Devers, C. (2024). *Testing, characterization and analysis of Class F fly ash geopolymers under oilfield testing conditions*. Unaiversity of Oklahoma https://hdl.handle.net/11244/340538.

Dong, Z., Zhao, B., Ao, K., Hou, W., Wu, L., Lu, H., & Shi, L. (2024). Review of evaluation methods for cement-formation interfacial bonding strength. In *Proceedings of the International Field Exploration and Development Conference, IFEDC 2023*. Singapore: Springer https://doi.org/10.1007/978-981-97-0256-5_42.

Drilling Manual. (2021). *Gas & oil well cement properties*. Drilling Manual Retrieved from https://www.drillingmanual.com/oil-well-cement-properties/ .

Drilling Manual. (2022). *Oil well cement testing simple guide*. Drilling Manual Retrieved from https://www.drillingmanual.com/oil-well-cement-testing-simple-guide/ .

Evans, G., & Carter, L. (1962). A study of cement-pipe bonding. *Journal of Petroleum Technology, 16*(02), 157–160. https://doi.org/10.2118/764-PA.

Gomado, F., Khalifeh, M., Kamali, M., Saasen, A., & Aasen, J. (2022). Sealing performance of geopolymer for plugging and abandonment; apple-to-apple scenario. In *SPE Norway Subsurface Conference, Bergen, Norway* https://doi.org/10.2118/209552-MS.

Helmy, Y., & Fakher, S. (2024). Evaluating the performance of Class F fly ash compared to class g cement for hydrocarbon wells cementing: An experimental investigation. *Materials, 17*(11), 2710. https://doi.org/10.3390/ma17112710.

Kamali, M., Khalifeh, M., Saasen, A., Godøy, R., & Delabroy, L. (2021). Alternative setting materials for primary cementing and zonal isolation – Laboratory evaluation of rheological and mechanical properties. *Journal of Petroleum Science and Engineering, 201*, 108455. https://doi.org/10.1016/j.petrol.2021.108455.

Kanjirakat, A., Carvero, A., Abraham, J. J., Abughaush, L., & Amani, M. (2026). High-frequency sonication as an unconventional solution to control fluid loss in water-based drilling muds. *ACS Omega, 11*, 7063–7073. https://doi.org/10.1021/acsomega.5c05870.

Khalifeh, M., Hodne, H., Saasen, A., Dziekonski, M., & Brown, C. (2018). Bond strength between different casing materials and cement. In *Paper presented at the SPE Norway One Day Seminar, Bergen, Norway* April 2018 https://doi.org/10.2118/191322-MS .

Ladva, H. K. J., Craster, B., Jones, T. G. J., Goldsmith, G., & Scott, D. (2005). The cement-to-formation interface in zonal isolation. *SPE Drilling & Completion, 20*(2005), 186–197. https://doi.org/10.2118/88016-PA.

Lavrov, A., Bhuiyan, M., & Stroisz, A. (2019). Push-out test: Why bother? *Journal of Petroleum Science and Engineering, 172*, 297–302. https://doi.org/10.1016/j.petrol.2018.09.067.

Liu, X., Nair, S., Cowan, M., & Oort, E. (2015). A novel method to evaluate cement-shale bond strength. In *SPE International Symposium on Oilfield Chemistry, The Woodlands, Texas* https://doi.org/10.2118/173802-MS.

Liu, K., Zhao, S., Li, W., Wang, H., Yi, D., Wang, J., & Guo, X. (2018). Research and application on improving the cementing quality of second-interface. *Oilfield Chemistry, 35*(1), 16–21. http://dx.doi.org/10.19346/j.cnki.1000-4092.2018.01.004.

Michel, F., & Courard, L. (2014). Particle size distribution of limestone fillers: Granulometry and specific surface area investigations. *Particulate Science and Technology, 32*. https://doi.org/10.1080/02726351.2013.873503.

Nasser, J., Jesil, A., Mohiuddin, T., Ruqeshi, M., Devi, G., & Mohataram, S. (2013). Experimental investigation of drilling fluid performance as nanoparticles. *World Journal of Nano Science and Engineering, 3*, 57–61. https://doi.org/10.4236/wjnse.2013.33008.

Peterson, B. (1963). Bond of cement compositions for cementing wells. *6th World Petroleum Congress* (pp. WPC–10123). Germany: Frankfurt am Main.

Siratovich, P., Heap, M. J., Villeneuve, M., Cole, J., & Reuschlé, T. (2014). Physical property relationships of the Rotokawa Andesite, a significant geothermal reservoir rock in the Taupo Volcanic Zone, New Zealand. *Geothermal Energy, 1*. https://doi.org/10.1186/s40517-014-0010-4.

Wagh, A. S. (2016). Chapter 16 – Chemically bonded phosphate ceramic borehole sealant. *Chemically bonded phosphate ceramics* (2nd ed., pp. 233–254). Elsevier 2016. https://doi.org/10.1016/B978-0-08-100380-0.00016-6 .

Wang, B., Feng, H., Huang, H., Guo, A., Zheng, Y., & Wang, Y. (2023). Bonding properties between fly ash/slag-based engineering geopolymer composites and concrete. *Materials, 16*(12), 4232. https://doi.org/10.3390/ma16124232.

Yan, X., Lan, H., Moore, I., & Ma, B. (2021). Flow properties of fresh mud (drilling fluid) used in horizontal directional drilling. *Journal of Pipeline Systems Engineering and Practice, 12*, 04020063. https://doi.org/10.1061/(ASCE)PS.1949-1204.0000517.

Yang, H., Fu, Q., Wu, J., Qu, L., Xiong, D., & Liu, Y. (2020). Experimental study of shear and hydraulic bonding strength between casing and cement under complex temperature and pressure conditions. *Royal Society Open Science, 7*(4). https://doi.org/10.1098/rsos.192115.

CHAPTER

4

Sustained casing pressures—Prevention monitoring and remediation

Rahman Ashena[a], *Ali Ghalambor*[b], *and Matteo Loizzo*[c]

[a] **Smart Energy Data Analytics (SEDA)-Group, Houston, TX, United States,** [b] **Oil Center Research, L.L.C., Lafayette, LA, United States,** [c] **Independent Consultant, Germany**

4.1 Introduction

Annular or casing pressure (CP) refers to the pressure detected in the annular space of a well's casings at the surface. The CP was divided into annular pressure buildup (APB) and sustained CP (SCP). The APB, sometimes called trapped casing or annular pressure by Zhang et al. (2019), is the term used when CP is only because of the natural effect of thermal expansion of annular fluid; this pressure may be bled off by opening the annulus side valves, and the pressure would not revert/build up. Note that this term is not frequently used by many practitioners.

On the other hand the (unplanned) SCP refers to the persistent pressure found in the annular space of a well's casings, even when the well is not actively producing or injecting fluids. In other words, SCP is the pressure that remains trapped between two casing strings or between the production casing and the completion string, with the pressure occurring because of well integrity failure at wellhead, casing, or cement allowing fluid migration. Thus the presence of SCP is very often an indication of a flowpath of hydrocarbons to the surface, even though it is not necessarily a sign of a leak outside of the well envelope (i.e., it is a failure of the primary containment barrier, but not the second). The SCP events have caused many wells worldwide to either undergo well control incidents (if

Well Cement Integrity
DOI: https://doi.org/10.1016/B978-0-443-51564-4.00011-4

the secondary barrier suddenly fails) or production halt and the associated necessity of workover or premature abandonment (if controlled, but with an unacceptable leak rate).

For efficient well integrity, SCP should not exist (ideally zero) or be minimal and constant in a well's active lifecycle: (1) after drilling a well and (2) during the production phase. After plug and abandonment (prior to reclamation), almost no jurisdiction would accept any SCP. However, the presence of SCP over time is the result of failure or lack of underground well integrity barriers, such as casing leaks, cement failures, or improper sealing. Indeed, it is one of the first signs that something is wrong with the well's underground barriers, that is, cement and casing barriers' sealing, and thus there would be only a single barrier left, that is, the outer casing and wellhead. SCP can be considered the final direct measurement of lack of zonal isolation. There are three hazards connected to SCP: peak pressure (if below MAASP), continuous flow rate, and hazards of the fluid (flammability, toxicity, etc.). These hazards drive the severity of potential accidents and thus risk of SCP. Pressure is not the only hazard for SCP occurrence. The greater the SCP, the greater the risk.

SCP should be continuously measured and monitored in both onshore and offshore wells, including subsea installations. Although SCP may remain stable or constant, according to API (2010), CP is considered SCP if it rebuilds to at least the same level after being bled down. If SCP increases gradually over time, it could indicate a developing issue that may lead to more severe consequences if left unaddressed. Identifying the specific annulus where SCP occurs—between which two casings—can help identify potential pathways for fluid migration to the surface. Based on some seniors' experience in the Middle East, the majority of SCP events in wells (say +90%) were observed in the annulus between the production and intermediate casings, and the rest were observed in next annuli. Li et al. (2018) presents an example of SCPs in different annuli of an offshore case study well with the true vertical depth (TVD) of ~2120 m (6955 ft) after 30 years of production. The SCPs in the three annuli range from 305 to 1726 psi.

Note that there are other terms used in different contexts for SCP: sustained annulus pressure, or casing–casing annulus pressure. Regardless of what it is called, this pressure should be measured and taken seriously.

This chapter covers consequences of neglecting SCP, nomenclature for different annuli in a well, causes of SCP, factors affecting SCP, and prevention, monitoring, and remedial measures in SCP cases.

4.2 Consequences

The consequences of ignoring SCP are one or more of the following: loss of efficiency or even production halt (because of well shut-in), environmental pollution, or even blowouts. When SCP is not recognized, it can lead to underground blowouts as well. The consequences of ignoring (not measuring, monitoring, and handling) SCPs are briefed as

- **Safety concerns:**
  - *Environmental contamination*: Leaking fluids, especially oil and brine (and then gas), can contaminate soil, water, or air, leading to significant environmental hazards. Such leaks below API 14B limits are not considered blowouts. This event may be invisible and thus more neglected in subsea offshore when subsea SCP monitoring is ignored, letting potential gas leak to the seabed, which would lead to significant damage to the marine environment.
  - *Blowout risk*: If SCP builds up unchecked, it could eventually lead to serious damage, rupture of the wellhead, and blowout at the surface where pressurized fluids are released uncontrollably to the surface. Or alternatively, it can cause fracturing of the formation at the casing shoe, causing an underground blowout with fluids either channeling the formations to penetrate and reach the next outer annulus or even reaching the surface (if exposed to shallow formations). These blowouts are the two most dangerous scenarios in well operations. An example of underground blowouts during operation because of lack of cement integrity and ignoring SCP was presented by Ashena et al. (2011). No proper and frequent monitoring of the SCPs was made during production of the land well in the Middle East.
- **Regulatory compliance:**
  Many regulatory bodies worldwide require operators to monitor and report SCP levels. Failure to manage SCP can lead to noncompliance with well safety standards and may incur fines or operational shutdowns.
- **Operational costs:**
  Remediation of SCP-related problems, such as repairing or recementing the casing, can be expensive and time-consuming. If the issue is severe, it could even lead to well abandonment, representing a significant financial loss for the operator/operating company.

Therefore prevention/mitigation of SCP means preventing or mitigating hazardous situations, controlling operational costs, and ensuring compliance with regulatory standards. Next, monitoring for proper detection of SCP-related problems is essential for the long-term success and safety of oil and gas or geothermal operations. It is essential to monitor the well's SCPs frequently, or better continuously using automatic systems.

4.3 Nomenclature for annulus

To properly name the various annuli in a well, refer to Fig. 4.1, which illustrates a well schematic with different annuli labeled in alphabetical

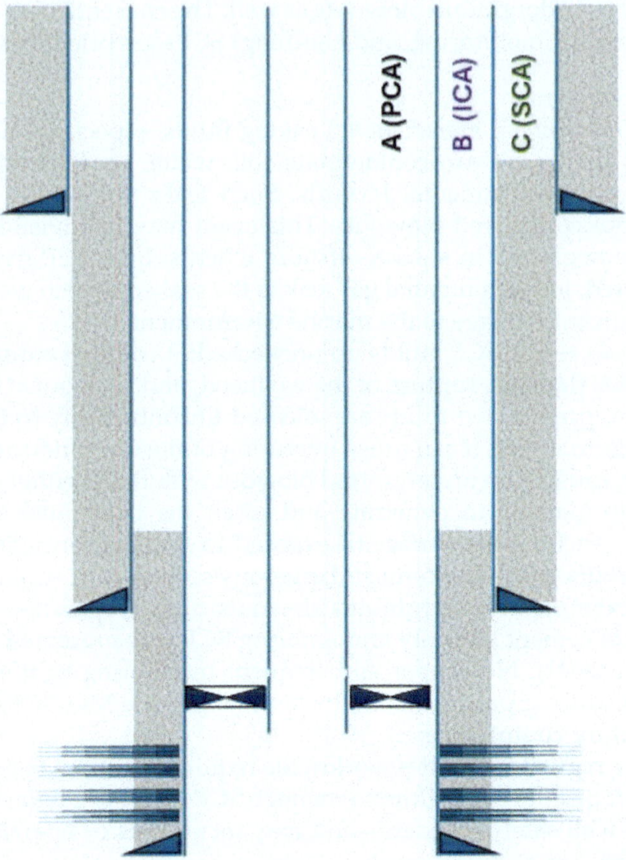

FIGURE 4.1 Different wellbore annuli named from A-annulus (between the completion string and production casing) to B-annulus (between the production and innermost intermediate casings), and C-annulus., and necessity of sealing them during production and after P&A. *ICA*, Intermediate casing annulus; *PCA*, production casing annulus; *SCA*, surface casing annulus. *(Redrawn from Li, X., Wang, C., Wang, Y., Li, F., & Xu, H. (2018). Practice of sustained casing pressure removal for JZ gasfield production well plug and abandonment in Bohai Bay. In Paper presented at the SPE Symposium: Decommissioning and Abandonment, Kuala Lumpur, Malaysia December 2018. https://doi.org/10.2118/193944-MS).*

order, starting with "A" for the innermost annulus. Therefore the A-annulus refers to the space between the completion string and the production casing, whereas B-annulus represents the area between the production and innermost intermediate casing. Subsequent annuli (C, D, etc.) are labeled accordingly, each representing the spaces between different casings within the well. It is important to note that this nomenclature system follows the reverse order compared with the driller's sequence of casing installation and cementing in the well. Note that the annuli behind the liners are not named alphabetically because they do not extend up to the

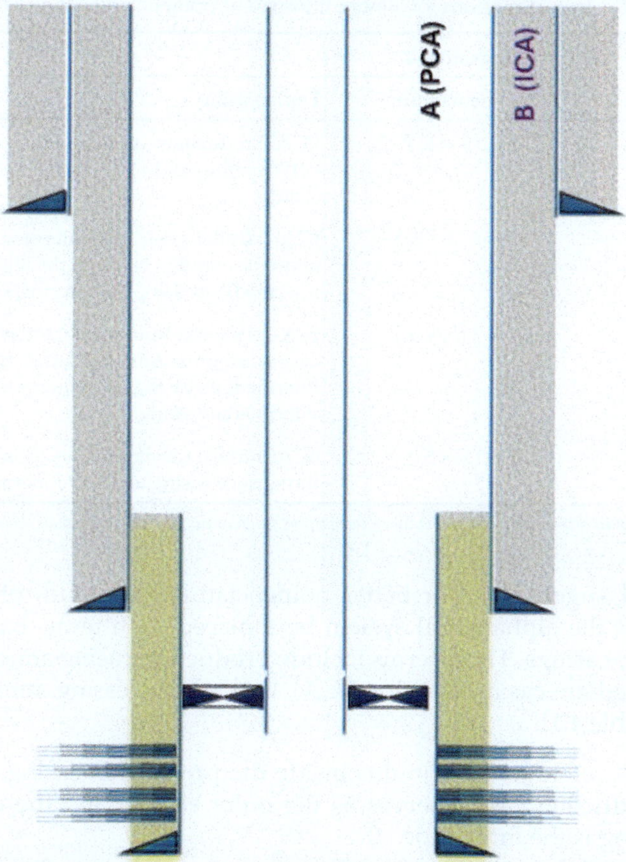

FIGURE 4.2 Annuli of a well with liner string. *ICA*, Intermediate casing annulus; *PCA*, production casing annulus.

wellhead. Instead, they are referred to by the liner size. For example, the *yellow shade* shows the "annulus behind the 7-in liner," as shown in Fig. 4.2. Although this nomenclature is commonly used by many companies (not all), it can lead to confusion for well engineers, as they often waste time trying to identify the correct annulus.

To avoid confusion, we present the following nomenclature suggestions in this chapter:

- First suggestion: Well's annuli should be named consistently by specifying the two sizes of the inner and outer casings or pipes, listed from the innermost to the outermost. For example, the B-annulus should be labeled as "B-Annulus (9 5/8 × 13 3/8)." This approach provides greater clarity and reduces confusion, as it would make it clear what annulus is referred to and keeps the typical 'industry utilized' nomenclature.

TABLE 4.1 Two suggestions for a more transparent nomenclature of a well's annuli.

	Suggestion-2	
Suggestion-1	**Nomenclature**	**Explanation**
A ($S_i \times S_o$)	PCA ($S_i \times S_o$)	PCA: Production casing annulus: annulus between the production casing and completion string
B ($S_i \times S_o$)	ICA-1 ($S_i \times S_o$)	ICA-1: (Innermost/first) intermediate casing annulus: annulus between the innermost intermediate casing and production string
C ($S_i \times S_o$)	ICA-2 ($S_i \times S_o$)	ICA-2: (Second from inside) intermediate casing annulus: Annulus between the second intermediate casing and innermost intermediate string
D ($S_i \times S_o$)	SCA ($S_i \times S_o$)	SCA: surface casing annulus: Annulus between the surface casing and 2nd-innermost string

S_i *is the (nominal) size or diameter of the inner casing/pipe in an annulus;* S_o *is the size or diameter of the outer casing.*

- Second suggestion: For better nomenclature for annuli of wells, we suggest the alphabetical system be replaced with terms based on the specific casings. These terms include production casing annulus (PCA), intermediate casing annulus (ICA), and surface casing annulus (SCA) (see Table 4.1).
  - PCA refers to the annulus inside the production casing, where the production casing serves as the outer casing, and the completion string is the inner pipe.
  - ICA refers to the annulus where the intermediate casing is the outer casing and the production casing is the inner casing. If there are multiple intermediate casings, ICA-1 refers to the annulus of the innermost intermediate casing, whereas ICA-2 refers to the annulus inside the second innermost intermediate casing, and so on.
  - SCA refers to the annulus between the surface casing and the outermost intermediate casing.

4.4 Causes of sustained casing pressure

Generally the same factors that contribute to cement and casing degradation and failure cause SCP, such as increasing microannuli at the cement-casing or cement-formation interfaces, damage to primary cement because of mechanical shock impacts during tripping, thermal cracking caused by different expansion properties of cement and casings, the condensation of hydrocarbons in microfractures, and chemical dissolution of cement in acidic formation brines (e.g., the reaction of CaO with CO_2 and H_2O giving

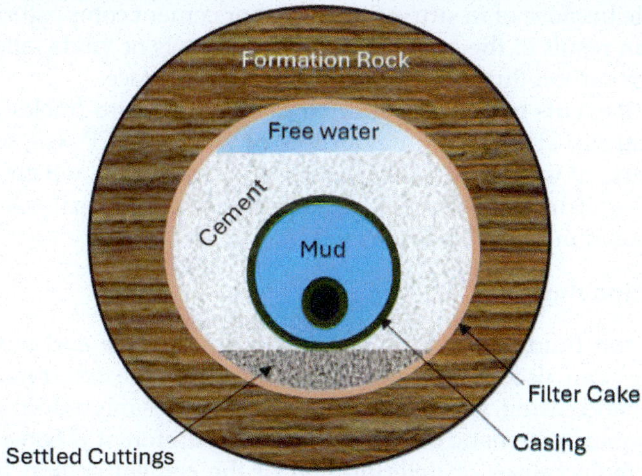

FIGURE 4.3 Potential challenges of cementing in high-angle wells: eccentric casing and logging tool as well as settled cuttings and free-water.

$CaCO_3$ and OH^- ions). There are many case studies of initially problem-free wells, which developed SCP over time (Bourgoyne et al., 1999). More specifically, the common causes for SCP are as follows:

Casing or completion string leaks:

- Over time the well casing might develop cracks or leaks. These defects can allow formation fluids (gas, oil, or water) to escape from one zone to another, accumulating pressure in the annular space.
- The completion string may leak over time, leading to fluid leaking to the annulus of the completion string and production casing. An unplanned SCP at A-annulus is usually a result of a leak in the primary barrier envelope, including the packer and completion string.
- Typically, production fluid flows through the completion string or tubing. However, in certain exceptional cases, such as with older or unconventional completions, production may occur through the A-annulus (Ashena et al., 2021). In these instances, flowing pressures would be present in the annulus.

Cement issues:

- Poor cement jobs are a common contributing factor to SCP. The cement, which is responsible for bonding and sealing the space between casing and formation, may not be properly placed during cementing because of factors such as geometry, bad design and execution, and the hydration effect. For instance, see Fig. 4.3 showing some cementing challenges in high-angle or horizontal wells from eccentric casing and logging tool as well as settled cuttings and free water. Over time, it may also fail or degrade because of in situ stresses or poor cement composition choices. This can result in the creation of microchannels or voids, allowing gas or hydrocarbon fluids to migrate toward the surface.

- The SCP events have been reported more after hydro-fracking stimulation (Landry et al., 2015). Regardless of whether SCP was seen before fracturing, it was almost always seen after fracturing, which shows the possible detrimental effect of fracking on cement integrity. An example was shown in Cana-Woodford field.

Formation fluids:

- Formation fluids are typically under high pressure and temperature, and they can also be corrosive. This can lead to the deterioration of the cement and casing, which, in turn, may allow gas migration; see Fig. 4.4 for a typical schematic of gas migration. Therefore it is considered one of the main causes of deterioration of the cement behind this casing and SCP. In a research study on 8122 wells in the outer continental shelf (OCS) made by Bourgoyne et al. (1997), 50% of the casing strings exhibiting SCP are in production casing. Next, 30% of the casing strings exhibiting SCP are in the surface casing (see Fig. 4.5).

Shallow-gas:

- Shallow gas can cause charging of the annulus between casing, pressure build-up, and development of SCP. In a Malaysian offshore field, there is a high probability of shallow-gas presence, which has caused 12 wells (60% wells in the field) experiencing SCP issues in their annulus. The SCP readings in these wells vary from 100 to 600 psi. The continuous increase in SCP has raised an alarm to the operator, necessitating production stoppage and well intervention (Ezani et al., 2021). In another case study related to SCP during drilling a Middle Eastern oil well (Nabaei et al., 2011), after running and cementing the intermediate casing to the depth of 613 m/2011 ft (where there was apparently an unexpected gas pocket), gas bubbles were observed at the cellar. The kick occurred because of gas migration in the light cement. Despite an attempt to control the flow, gas pressure created induced fractures in the shallow formations, spreading all around the drilling area. Upon the gas reaching the electric generators, an explosion happened with fire flame of about 70 m (~230 ft) high.

Induced fracturing:

- Because of the wrong selection of cement density, induced fracturing of underground formations can lead to cement losses. Thus the resultant cement sheath cannot reach the required top of cement (TOC), making the well susceptible to formation fluid intrusion into the annulus.

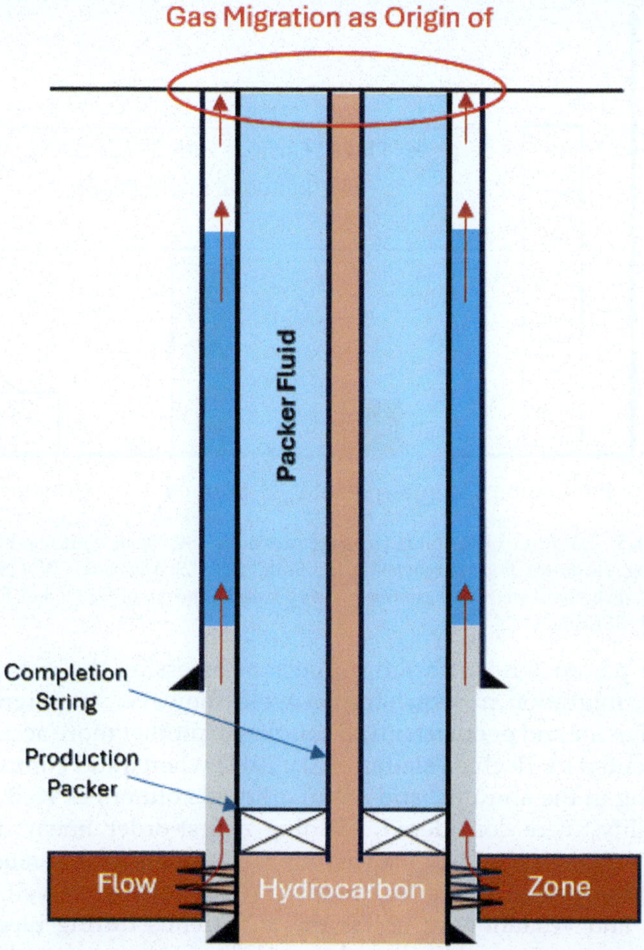

FIGURE 4.4 Gas migration through the cement. *(Redrawn from Kamgang, S., Hanif, A., Neupane, R., & Matthew, G. (2023). Cement Bond Logging; A Generational Review of Technologies & Log Interpretation. Paper presented at the ADIPEC, Abu Dhabi, UAE, October 2023. https://doi.org/10.2118/216898-MS).*

Faulty wellhead equipment:

- Leaks or malfunctions in wellhead components, like valves or seals, can cause pressure to accumulate in the annular space.

4.5 Factors affecting sustained casing pressure

Fig. 4.6 shows a schematic of parameters involved in SCP creation when TOC is not on the surface. Huerta et al. (2009) provided a simple model for anticipating SCP in four cases using some assumptions for flow geometry

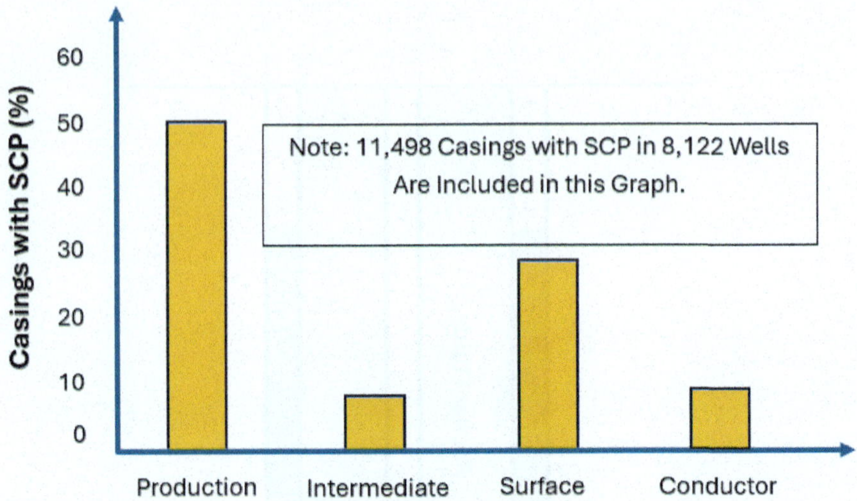

FIGURE 4.5 Location of sustained casing pressures (SCPs) in combined database by casing string. *(Redrawn from Burgoyne, A. T., Scott, S. L., & Manowski, W. (1997). Areview of sustained casing pressure. Outer continental shelf's wells. Bureau of Safety and Environmental Enforcement (BSEE)).*

and flow parameters, with the purpose of assessing the risk associated with CO_2 migration in sequestration wells; however, they ignored CO_2 phase behavior and geomechanical coupling. Another more accurate work was presented by Rocha-Valadez et al. (2014) where mud compressibility, gas leaking in the annulus, and annular liquid column, as well as cement permeability, were considered, yielding a first-order linear differential equation to be solved. They showed the SCP stabilization versus time.

In addition, the SCP profiles/trends may be modeled as well. In a work by Kinik and Wojtanowicz (2011) the SCP trends during bleed-off and build-up tests were analyzed, and a model was presented to benchmark in a case study.

There are several factors affecting SCP. Zeng et al. (2018) discussed effects of several factors influencing the stabilized time, including damaged cement permeability, wellhead temperature, and deformation of annular space.

Four main factors affecting SCP, investigated in the literature, are as follows:

- Gas source depth:
  Su et al. (2021) shows that the deeper the gas source is, the greater the SCP.
- TOC depth:
  If the TOC is above the previous casing shoe, this can entrap drilling fluids in the annulus. High-temperature fluid will heat the casing and cement during well testing and production, which will induce thermal

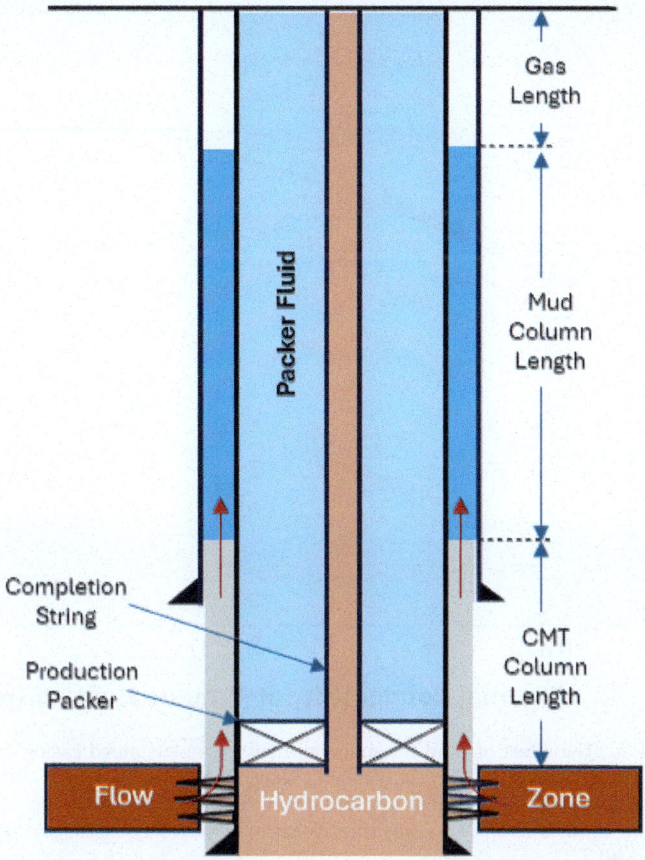

FIGURE 4.6 Schematic of dimensions and parameters in a cement/mud annular system. *(Redrawn from Burgoyne, A. T., Scott, S. L., & Manowski, W. (1997). Areview of sustained casing pressure. Outer continental shelf's wells. Bureau of Safety and Environmental Enforcement (BSEE)).*

stress and SCP in the annulus. Wang et al. (2020) conducted some modeling with sensitivity analysis of changing TOC only in a single annulus and found an increasing SCP with decreasing TOC. The finding can be justified as the shallower the TOC is, the greater the SCP would be (usually). The shallower TOC is in an annulus, the greater the bottom-hole temperature and the greater the fluid expansion effect, and thus the greater the SCP is expected to be (assuming other parameters are constant). This effect may also be attributed to greater possibility of gas migration through a thinner cement sheath in the annulus to the surface. However, in a more comprehensive work by Wang et al. (2022), they showed this is not necessarily the case, and it is case sensitive.

- Annular fluid density:
  The lighter the annulus fluid is, the greater the SCP. This was confirmed by another work by Su et al. (2021). The effect of annulus fluid density

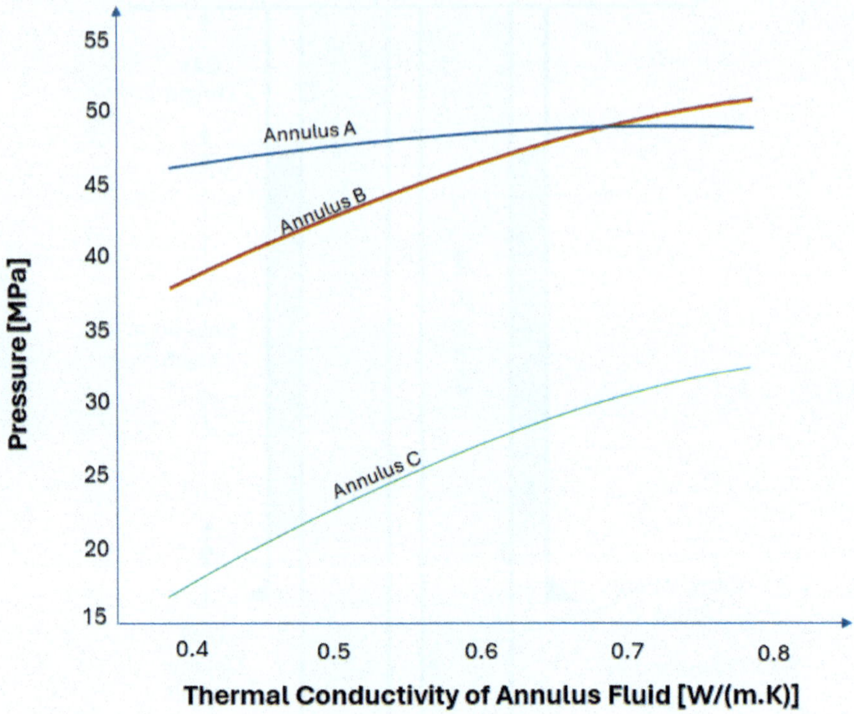

FIGURE 4.7 The effect of fluid thermal conductivity on sustained casing pressure. *(Redrawn from Wang, Y., Xin, S., Gao, D., Wang, J., Ning, B., & Zhao, Z. (2022). Sustained casing pressure: One of the most important factors determining the integrity of deepwater wells. In Paper presented at the Offshore Technology Conference, Houston, Texas, USA May 2022. https://doi.org/10.4043/31759-MS).*

on magnitude of SCP was simulated in a case study, with the vertical well depth being at 4120 m (~13,517 ft) and the cement top at 2320 m (~7612 ft).

- Thermal conductivity:
  The greater the thermal conductivity of the annular fluid in Fig. 4.7, the greater the bottom-hole temperature, and thus the greater the thermal effect on expansion and the greater the SCP.

4.6 Control measures

There are some essential measures to be taken to deal with the SCP phenomenon:

✓ Proper design
✓ Monitoring
✓ Remediation

These measures are explained as follows:

4.6.1 Prevention or mitigation by recommended cementing design and practices

4.6.1.1 Prejob simulation

Landry et al. (2015) provided several case studies in Cana-Woodford shale field to distinguish failure points and provided recommendations on improving cementing operations with 95% success rate in cement integrity of wells drilled from 2013 to 2015. One of the key points of the study was the advantage of cementing simulation before the actual operation; thus, as per API (2010), it is recommended to use modeling or simulation tools to predict cement performance desired under expected conditions against the observed failure mechanisms. The software may be trained using information from past wells' cementing jobs, including lab mechanical tests. The software should be able to conduct triaxial stress analysis of cement using mechanical lab tests' properties such as uniaxial compressive strength, tensile strength, and Poisson's ratio to see how the cement is expected to perform when subjected to bottom-hole temperature and pressure/in situ stresses, which may be thermo-poro-elastically modeled using equations in Ashena et al. (2018, 2019). This would be more economically viable than trial-and-error methods.

4.6.1.2 Cement properties and placement design

To prevent bad cement quality and later SCP, it is recommended to use carefully designed cement properties and formulation and placement. When selecting cement slurry density, it is advisable to first conduct a geomechanical analysis to ensure accurate selection, especially in formations prone to fracturing or in high-pressure high-temperature (HPHT) conditions. If the density of the cement slurry prepared at the rig site is lower than the design value, it could result in higher cement free-water levels, longer thickening time, and lower acoustic impedance compared with the theoretical values. Changes in cement density can be more clearly observed using an ultrasonic cement bond log (CBL), where the acoustic impedance is measured.

There are two important points related to cement thickening. First, during the thickening time cement slurries usually have a transition time before they set/harden (from the liquid to solid state), behaving like a gel with its hydrostatic column pressure declining, which may be even lower than the formation pore pressure (underbalance) allowing formation fluid flow or gas invasion (Ashena, 2025). Extended slurry thickening time may exist because of excessive safety margins for placement time and unsuitable pump displacement rates. In the investigation of causes for "deep" drilling loss of well control (LOWC) events in the US GoM OCS and the regulated areas, Holand (2017) mentioned that from 2000 to 2015 four LOWC events occurred while the cement was setting after running casings. The blowout database shows that this has also been a problem in the 1980s and 1990s. This type of primary barrier failure is common for shallow LOWC events as well. To address this limitation, first, thickening time should not be longer than necessary, and a reduction of the transition time is suggested. Next, though it is not a common practice, it is recommended to apply pressure in the annulus for sufficient time to maintain the hydrostatic slurry pressure above formation pore pressure, preventing fluid migration and loss of zonal isolation. The second point is that too short thickening time will cause cement premature setting, which is a cause of discrepancy between the actual and theoretical TOCs. Thickening time should be designed carefully such that pumping and cementing operation should stop upon reaching this time.

At lab environment, following the hardening, cements are expected to show in the lab large enough shear bond strength and hydraulic bond strength, high strength and flexibility, and low permeability. In geothermal or HPHT wells, it is important to analyze the effect of pressure and temperature (particularly temperature) on cement properties, especially its bond to rock and casing pipe, using software as well as lab experiments.

Because of its high durability, neat cement is often used as the tail cement in two-stage cementing, typically placed across the shoe track. Neat cement is defined in the industry as competent cement without additives. When selecting neat cement, it is crucial to ensure that the formation pressures are within safe limits to prevent fracturing the formation. For instance, the neat Class G and H cement slurries have densities of about 15.8 and 16.2 ppg. It is advised not to use additives in the tail cement where the shoe track would be covered, because additives may make the cement susceptible or less durable with time when exposed to downhole stresses and pressure and temperature conditions.

It is known that cement bulk shrinkage of up to 5% can occur because of cement hydration reaction (Reddy et al., 2009). To mitigate cement shrinkage and its effect on debonding, it is recommended to add a postset expanding additive to the cement slurry.

In special cases such as cementing long casing or liner strings or wells with shallow gas, the two-stage cementing method may be recommended to improve cement slurry placement. Nafikova et al. (2021) considered long cement columns to be in the range of 1300–1500 m (4265–4921 ft). Fig. 4.8 shows an example of two-stage cementing for the shallow gas. Nowadays, foam-cementing is also becoming a trend.

It is vital that the cement properties are optimized based on results of a downhole stress model to sustain higher levels of mechanical stress. Although it is challenging to totally avoid a mechanical failure in the very long term, recent and ongoing research has focused on achieving this goal. To prevent mechanical failure, it is important to add the self-healing function to cement, which is designed to activate if microannuli or tiny cracks start to develop in the cement. If such microannuli or cracks

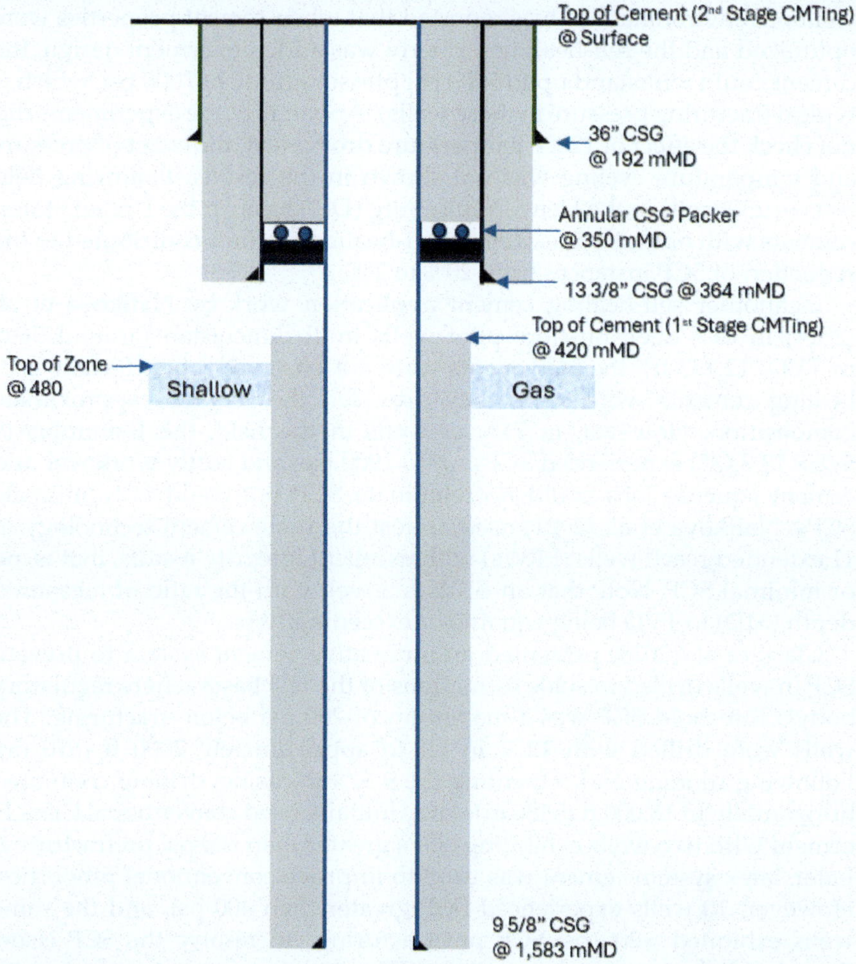

FIGURE 4.8 Two-stage cementing method applied in an offshore case study to prevent sustained casing pressure in future wells. *(Redrawn from Ezani, F. S., Thuzar, M., Kumar, A. K., & Lau, C. H. (2021). Prevention of historically challenging sustained casing pressure from shallow gas through re-engineered cement design and execution methodology. In Paper presented at the International Petroleum Technology Conference, Virtual March 2021. https://doi.org/10.2523/IPTC-21429-MS).*

extend, they can develop into microchannels through which hydrocarbons can flow, eventually reaching the surface. Therefore the system has the capability to repair itself and restore the integrity of the cement sheath. Langley et al. (2018) investigated self-healing cements with simulation and lab studies. They conducted simulations on several cement formulations, including Class G and optimized self-healing cements of 12 lb/gal, after curing cement for 72 hours at 230°F and 3900 psi to properly simulate the wellbore environment. It was reported that when cement properties were optimized and the self-healing property was added to cement design, the cement could withstand up to 5000 psi pressure (but not 7000 psi, which is typical fracturing pressure in these wells). Apparently, the experiments did not check the effect of 120°F temperature drop. Next, the effect of pressure and temperature cycling was not shown in the results. Following field tests in 250 wells in the Denver–Julesburg (D–J) basin in the United States, cements with optimized as well as self-healing features contributed to the reduction of SCP instances from 20% to 2%.

In another self-healing cement application work by Nafikova et al. (2019) in two deep offshore pilot wells in Turkmenistan (around 3698 m-TVD/12,133 ft), the SCP events were not observed when flexible self-healing cements were used; they provided the cement's approximate composition. However, in 71% of wells in the field, the B-annulus (9 5/8 × 13 3/8″) experienced SCPs up to 2900 psi, and costly workover and cement squeeze jobs could not eliminate SCP but could only mitigate SCPs. Nafikova et al. (2021) used almost the same cement technology in 11 extended reach wells (ERWs) with excellent integrity results, that is, no or minimal SCP. Note that an ERW is a well with the ratio of measured depth (MD) to TVD being equal to or exceeding two.

Clark et al. (2023) presented an innovative cement system to prevent SCP in wells drilled in shale formations of the D–J basin where regulatory body's threshold SCP was a maximum of 200 psi before fracturing. The wells were drilled with 12 ¼-in bit to approximately 2000 ft (610 m). Following running and cementing the 9 5/8-in casing, drilling continued horizontally to 18,000 ft (5486 m). They initially used conventional Class H cement with two wells exhibiting SCPs greater than 600 psi prefracturing. Later, latex system cement was used to improve conventional properties. However, 10 wells experienced SCP greater than 400 psi, and the same wells exhibited >200 psi SCP postfracturing. To resolve the SCP issue, they decided to use an innovative SCP mitigation cement system (with composition not revealed) for cementing part of the annulus behind 9 5/8-in casing. In the field, just few SCP events were observed (only 1% of wells for prefrac and 5% for postfrac), and CBLs showed good cement quality. Note that the final cement properties at lab were improved in some respects: the transition time was kept almost the same at 34 minutes (not a good sign of improvement), 40% greater shear bond strength and 75% lower permeability than a conventional Class H cement. The greatest permeability of the innovative cement sheath was related to the 14 ppg cement with 25.6 μD. As another good feature, the cement fluid loss was 30% lower with 24 cm^3/minute. In terms of strength, the 24-hour cement compressive strength (UCS) was 949 psi, 60% greater than that of the base cement. On the other hand, the 72-hour cement UCS was 1241 psi, 4.5% lower than the latex cement system.

Chapter 4 mentions some good cement placement practices to ensure good cement quality/bond. Before cementing, extreme care should be taken to avoid a long rat-hole with lighter mud than cement slurry at the bottom of the wellbore to prevent possible contamination of cement at the shoe track following cementing. The rat-hole refers to the interval between the bottom of the casing pipe and the bottom, and such a completion type is called "off-bottom." One of the causes of the Macondo blowout in 2010 was the "rathole" of 54 ft (16.5 m) filled with mud of 14.17 ppg, whereas the cement density was 16.7 ppg; that is, ~2.5 ppg lighter mud in the rat hole (BP, 2010).

Proper centralization of the casing using enough rigid centralizers in the open hole and cased hole is required to ensure a uniform cement sheath around the casing pipe at all depths. Centralization is particularly important to be strictly followed in horizontal wells (API, 2010; Ghalambor, et al., 2000; Landry et al., 2015). To enable an increased number of centralizers in an extended-reach horizontal well with minimal risk of stuck casing during a run, rigid centralizers with low coefficient of friction and high side-loading capability were used along the open-hole section (Nafikova et al., 2021).

4.6.1.3 Displacement

A critical problem leading to poor cement bond and quality is ineffective washing of the drilling mud and mud cake off the wellbore wall while pumping the cement slurry behind the casing pipe. In this case the cement bond would most likely be poor. Maintaining sufficiently high slurry annular velocity during displacement is crucial for efficient mud removal, preventing cement contamination, cleaning excessive mud cake, and avoiding premature thickening. This is particularly important in horizontal or ERWs where displacing the mud is a big challenge. Nafikova et l. (2021) mentioned using a high displacement rate of 15–18 bbl/minute (630–756 gal/minute or gpm) for good mud displacement in an extended reach horizontal well with 3449 m/11.316 ft MD (around 1500 m/4921 TVD).

API (2010) just mentions the necessity of proper spacer design for good drilling fluid removal before cement is placed. Many methods or technologies have been developed for spacers. Some practitioners suggested using fiber technology in the spacer fluid for effective cleaning of non-acqueous muds on casing wall. This is important because one of the root causes of SCP is failure to remove all the nonaqueous mud from the annulus during cement placement, causing a large channel of formation fluid migration. Kumar et al. (2017) showed ultrasonic CBLs at depth 4000 m (13,124 ft) for the two cases, inferring that cement bond quality improves when an updated spacer (combination of optimized surfactant and fiber, right side) was used in lieu of conventional surfactant spacer (left side).

In another case study by the same authors, it was shown that a foamed spacer improved cementing jobs and mitigated the SCP in a case study well with shallow reservoir of 1250 m (4101 ft) in India. The good cement bond with minimal SCP was attributed to foam's good rheological properties, including its high yield point, contributing to its effectiveness in cleaning drilling mud, mud cake, and cuttings from the wellbore walls.

In fracture-prone zones, it is recommended to use a lost circulation material fluid as spacer ahead of cement slurry for the purpose of minimizing cement loss.

In shallow-gas cases or some special cases, it may be recommended to displace the cement slurry up to the surface and get (cement slurry) returns at the surface.

4.6.1.4 Cement bond evaluation

After wait on cement, it is recommended to run CBLs to recognize cement bond quality so that cement quality is evaluated. In case of bad cementing, remedial cementing may be required before starting production. In special cases like for the analysis of third interface echoes, it is recommended to use advanced ultrasonic CBL tools with flexural attenuation (Ashena et al., 2024; Landry et al., 2015). Examples of ultrasonic logs are isolation scanner or Epilogue (new generation of isolation scanner) taken by SLB. For more information on CBLs and evaluation, refer to Chapter 5.

4.6.1.5 MAASP and risk analysis

To prevent or mitigate SCP, it is critical to be aware of the critical or maximum allowable SCP. Therefore it is vital to determine the maximum annular surface pressure (MAASP), and not to let SCP exceed that value. The standard ISO (2017) provides formulas for determining the MAASP for the A, B, C, and other annuli, provided in SI units. The calculations for the A-annulus are more tedious than other annuli; this can be noticed by comparing the procedures in Appendices A and B at the end of the chapter.

In brief, there are two main criteria considered for finding MAASP:

a) *Maximum allowable wellhead operating pressure (MAWOP):* Both API and MMS management programs consider MAWOP as the criterion for the maximum allowable SCP. MAWOP provides a safety margin for failure scenario of collapsing of the inner tubular/pipe and bursting the outer casing pipe. The MAWOP is determined as the minimum of either of the following resistances:
   - ✓ 50% of outer casing's burst resistance,
   - ✓ 80% of the next outer casing's burst resistance, or
   - ✓ 75% of the inner casing's collapse resistance.

For the outermost annulus, the MAWOP is determined as the minimum of:

✓ 30% of the outer casing's burst resistance. A comparison reveals that the MAWOP of the outermost annulus is estimated to have a lower burst resistance than the inner annular layers.

✓ 75% of the inner pipe's collapse.

b) *Formation fracture pressure:* The second criterion for estimating the maximum SCP is based on the casing shoe fracture particularly when the formation is exposed in the annulus (cement did not extend up to cover the formation). For the calculation, the leak-off test data is required. Unfortunately, sometimes this criterion is ignored by some practitioners. Note that burst resistance is also called the minimum internal yield pressure in some contexts. Kinik and Wojtanowicz (2011) emphasized considering this criterion and integrating it with the previous criterion to develop a quantitative risk assessment model for SCP.

An example case study (Fig. 4.9) of maximum SCP is presented with the maximum allowable SCP calculation process being shown in Table 4.2. The bolded values show the maximum allowable SCPs, which should not be exceeded during operations, including the production phase. The case study shows that for the B-annulus (the deepest annulus), the critical SCP should be based on casing shoe pressure, whereas for the other annuli, the MAWOP calculations are the right ones. The calculations are not done for the A-annulus.

4.6.2 Monitoring

The MMS (2010)/BOEMRE 30 CFR 250 Annular Casing Pressure Management for Offshore Wells provides criteria for monitoring and testing of wells with SCP. In addition, the American Petroleum Institute's recommended practice-90 (API, 2006) gives guidelines for managing SCP and defines wells with different levels of environmental risks. The Norwegian standard NORSOK D-010 also defines monitoring requirements and tolerance criteria, based on API RP 14B.

Note that API Spec 17D forbids accessing outer annuli (like B or C annuli) in subsea wells. This standard hinders effective monitoring and management of SCP. Following the release of the API Technical Report 17TR3 in November 2004, titled "An Evaluation of the Risks and Benefits of Penetrations in Subsea Wellheads Below the BOP Stack," prepared by Young, K., and Ayers, R., API decided to prohibit access to outer annuli through the wellhead housing. Authors are not sure if this prohibition is still in place, though it is evident that operators worldwide continue to refuse access to annuli outside of the A-annulus. A revision of this standard may be necessary.

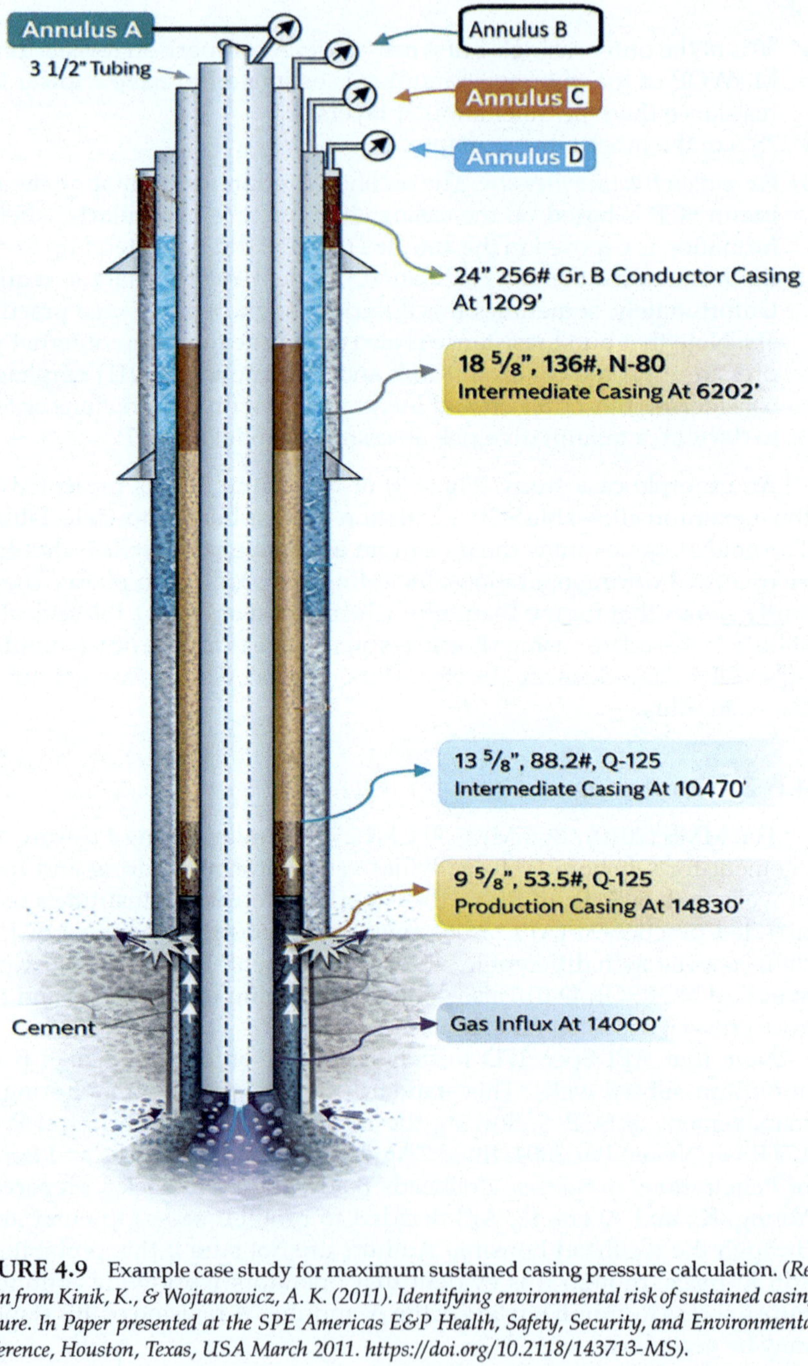

FIGURE 4.9 Example case study for maximum sustained casing pressure calculation. *(Redrawn from Kinik, K., & Wojtanowicz, A. K. (2011). Identifying environmental risk of sustained casing pressure. In Paper presented at the SPE Americas E&P Health, Safety, Security, and Environmental Conference, Houston, Texas, USA March 2011. https://doi.org/10.2118/143713-MS).*

TABLE 4.2 Comparison of MMS's maximum allowable wellhead operating pressure (MAWOP) criterion and critical sustained casing pressure (SCP) based on shoe fracture resistance.

Annulus	Outer CSG	Burst	Collapse	50% Burst	30% Burst	75% Collapse	80% Burst	MAWOP (MMS criterion)	Critical SCP, based on shoe fracture resistance
		(psig)							
Ann-A (not calculated)	9 5/8″ 53.5#, Q125	12,390	8440	NA	NA	6330	NA	NA	NA
Ann-B (9 5/8 × 13 3/8)	13 5/8″ 88.2# Q125	10,030	4800	5015	NA	3600	8024	Min (5015, 4168, 6330) = 4168	**1489**
Ann-C (13 5/8 × 18 5/8)	18 5/8″ 136# N80	5210	2480	2505	NA	1860	4168	Min (2505, 1276, 3600) = **1276**	1424
Ann-D (18 5/8 × 24)	24″ 246# Gr.B	1595	742	798	479	557	1276	Min (479, 1860) = **479**	558

(Data from Kinik, K. & Wojtanowicz, A. K. (2011). Identifying environmental risk of sustained casing pressure. In Paper presented at the SPE Americas E&P Health, Safety, Security, and Environmental Conference, Houston, Texas, USA March 2011. https://doi.org/10.2118/143713-MS).

Monitoring the SCP is carried out as follows:

4.6.2.1 *Regular monitoring*

The SCP is routinely monitored during well production, where the pressures in the annular spaces are recorded over time. This can help detect early signs of pressure buildup and potential integrity issues. It is essential that reliable (accurate) annulus pressure data is recorded on every well frequently.

Based on API (2006), it is required to monitor SCP, followed by bleed-off build-up test if the pressure becomes greater than 100 psig. Bleed-off should take for 24 hours through a ½-inch needle valve or until the CP reduces to zero. Operators are required to document and keep track of test records for at least 2 years. The types and amounts of fluids recovered during the bleed-off as well as the CP at the beginning of the test should be recorded. Following the bleed-off, the needle valve is closed to monitor pressure build-up for 24 hours.

If the pressure bleeds off to zero, and with no subsequent pressure increase during the build-up period, the annulus in question does not have SCP issue. Secondly, if the pressure bleeds to zero with build-up to the same or lower pressure (in 24 hours), the annulus is considered to have a tolerable leak rate, and barriers for pressure containment are adequate. Thirdly, if the pressure does not bleed to zero (in 24 hours), pressure containment barriers may have partially failed, and the leak rate may be unacceptable; see an example in Fig. 4.10.

Methods to monitor annular pressure:

4.6.2.2 *Traditional method*

To traditionally monitor the annulus pressure, a pressure gauge should be installed on an instrument flange, with either one or two gate valves between it and the wellhead side outlet/annulus (see Fig. 4.11A). Some operating companies do not prefer to permanently install annulus pressure gauges. Instead, often the field operator brings a freshly calibrated gauge to the well-site and installs it on the wellhead each time that pressure measurement is called for. The reason for this preference is related to the fragile nature of gauges. But this would further add to labor work and operational risk. In either case, there are the following disadvantages of this method (Brodie, 2016):

1. The gauge often must be manually read. This introduces the risk of erroneous recording and possible negligence.
2. The grease that is used to maintain the gate valves often creates a blockage between the gate valve and the gauge, rendering the gauge data unreliable.

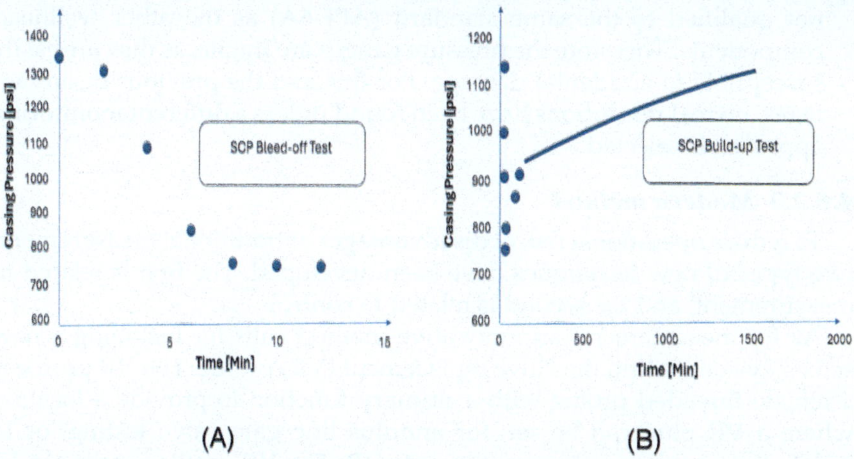

FIGURE 4.10 An example bleed-off build-up test showing unacceptable sustained casing pressure (SCP) issue with casing pressure not dropping or getting to zero after bleed-off, with SCP of 750 psi. *(Redrawn from Kinik, K. & Wojtanowicz, A. K. (2011). Identifying environmental risk of sustained casing pressure. In Paper presented at the SPE Americas E&P Health, Safety, Security, and Environmental Conference, Houston, Texas, USA March 2011. https://doi.org/10.2118/143713-MS).*

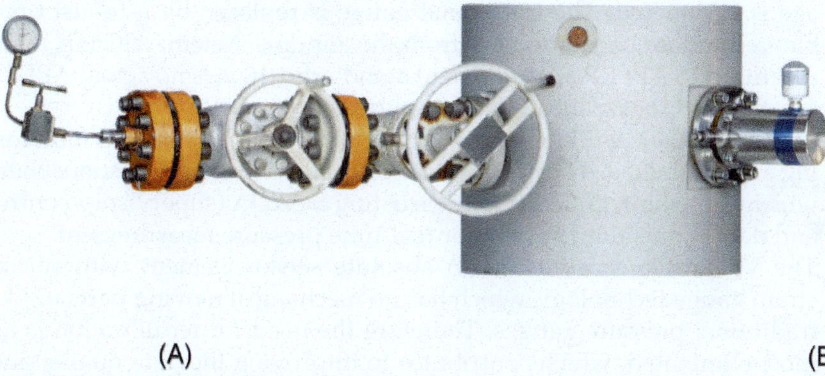

FIGURE 4.11 (A) Typical traditional annulus monitoring arrangement and (B) modern valve replacement plug sensor approach. *(Redrawn from Brodie, A. (2016). The autonomous management of wells with unplanned sustained casing pressure. In Paper presented at the Abu Dhabi International Petroleum Exhibition & Conference, Abu Dhabi, UAE November 2016. https://doi.org/10.2118/183077-MS).*

3. When the annulus line gate valves are opened to expose the pressure gauge to the annulus pressure, the barrier envelope is effectively extended to include the annulus line valves, instrument flange, capillary lines, and gauge itself. The gauges and capillary lines used are generally

not qualified to the same standard (API 6A) as the other wellhead components. Therefore the pressure gauges are fragile, as they are easily susceptible to accidental damage. For this and the previous disadvantages, no real resolutions have been found unless a different monitoring approach is selected.

4.6.2.3 Modern method

To reduce operational risk of disadvantages of traditional measurement, two types of new techniques have been developed. The first is related to measurement, and the second is related to control.

As for measurement, an innovative method calls for installing a new sensor system within the valve replacement (VR) profile. The VR profile is a female threaded profile with a primary function to provide a location where a VR plug can be set, for annulus line gate valve testing, or to isolate the annulus in the event of gate VR. The VR profile sensor can be best described as "Intelligent VR Plugs" (Brodie, 2016); see Fig. 4.11B for a schematic. The method has the following advantages:

1. The requirement for gate valves and an instrument flange between the sensor and the annulus is eliminated. This means that concern related to gate valve grease blocking the sensor is eliminated (main advantage).
2. The problem of fragile pressure gauges susceptible to accidental damage is eliminated. The traditional gauge is replaced by a robust double/redundant barrier to flow from the annulus system, which is manufactured to API RP 6A equivalent standards (there is no actual API RP 6A standard for wellhead data monitoring systems).
3. The manpower requirements and operational risk for manual monitoring are eliminated. The system automatically delivers continuous data, which can be integrated into the existing SCADA (supervisory control and data acquisition) system for real-time pressure measurement.
4. The VR-profile sensors use an absolute sensor element with silicon strain gauge technology, which has no mechanical moving parts unlike traditional pressure gauges. Therefore the need for recalibration is almost eliminated, which contributes to improving the data quality and reducing manual intervention requirement.

Regarding control and autonomous management of SCPs, a control system is designed using a programmable logic controller (PLC) to integrate the SCP monitoring system (VR sensor) with the surface annulus safety valve (SASV) hydraulic control system to vent/bleed-off gas when SCP becomes too high. The SASV is a new technological feature to allow opening or closing the well, which operates very similar to subsurface safety valve in the completion string. This valve, which was first developed for gas-lift wells, provides additional safety for well production. In this system the VR sensor continuously measures the annulus pressure and sends the data to

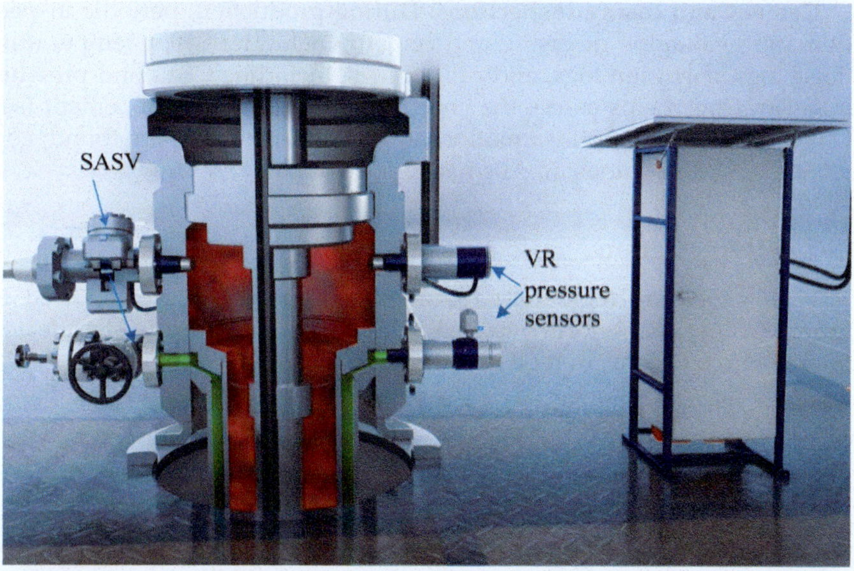

FIGURE 4.12 Schematic of regular monitoring by the integrated surface annulus safety valve (SASV) and valve replacement (VR) profile sensor programmable logic controller–controlled system. *(Redrawn from Brodie, A. (2016). The autonomous management of wells with unplanned sustained casing pressure. In Paper presented at the Abu Dhabi International Petroleum Exhibition & Conference, Abu Dhabi, UAE November 2016. https://doi.org/10.2118/183077-MS).*

the PLC. The PLC is programmed to act when the SCP exceeds a preset maximum value, either by notifying production operators or, ideally, by automatically bleeding off the SCP from the annulus. In the latter case, hydraulic pressure from a hydraulic accumulator system, controlled by the PLC, is used to open the SASV until the annulus pressure reaches a safe level. The bleed-off destination can be directed to closed drains (either on an offshore platform or an onshore well pad) or a production flowline (in conventional land operations). This automation system may also be used to maintain continuous production "under dispensation," as will be discussed in the "remediation" section (see Fig. 4.12 for a schematic).

4.6.2.4 Other related monitoring

The other related monitoring is as follows:

Wellhead maintenance: Regular maintenance of wellhead equipment (such as valves and seals) ensures that the wellhead does not suffer from mechanical failure so it can handle it safely in case an SCP event occurs. This includes checking for leaks and replacing damaged parts.

Cement and casing inspections: During production, periodic inspection and evaluation of casing and cement, including CBLs, temperature noise logs, corrosion logs, and caliper logs, including CBLs and pressure integrity tests, helps assess the condition of the casing and cement (see Chapters 5–8 for more information). If integrity problems are found, corrective remedial actions may be needed.

4.6.3 Remediation

With increasing regulatory requirements, there is a greater need to improve remedial performance and reduce costs. To address wells affected by SCP, three effective remedial actions are recommended:

✓ Bleed-off (or sometimes lubricate and bleed),
✓ workover operations, and
✓ premature abandonment.

In cases where SCP is considerably high but has stabilized, operators may use pressure relief valves or venting systems to release excess pressure and avoid the SCP exceeding the MAASP. It is sometimes called continuing production "under dispensation." In this case, increased manpower is needed for monitoring and manual bleed-off unless automation systems are used. Brodie (2016) suggests retrofitting the automation technologies originally used in gas lifting for autonomous annulus pressure monitoring and bleed-off without manual intervention (Brodie, 2016). The automation system was already explained in the previous section, "Regular Monitoring," and shown in Fig. 4.12. It is worth noting that, in some cases, the lubrication and bleed-off method may be used. This involves injecting heavy fluid into the annulus while bleeding off the gas. The advantage of this approach is that it helps reduce the risk of gas leaks from the annulus. However, as for the disadvantage, it is important to perform accurate calculations so that this method would not lead to the casing shoe fracture. McDonald and Li (2017) studied injection of a high-ratio sodium silicate (SiO_2 to $NaO = 4.5$) before remedial cement squeezing into the annulus of a trial well in Canada to shut off its low-rate vent flow. The results ranged from complete zonal isolation to reduced SCP; this method needs further research in chemistry and placement optimization.

In some extreme cases, SCP events may be resolved by resin injection. However, in some extreme cases where pressure becomes extremely high or it does not stabilize by venting (rising pressure) or resin injection, it is essential to cure the lack of well integrity by well intervention or workover operations (either rig-based or rig-less). The operation would usually include cement squeezing or repairing the casing. However, in other cases where workovers are not economically viable, the well may be prematurely abandoned.

As for workover, prior to starting the operation, it is important to diagnose the gas source by characterizing the gas properties (composition and isotope analysis). Thus fluid sampling from the annulus is a prerequisite for recognition of the zone from which the hydrocarbon or gas is rising.

Next, there can be guesses on the migration pathway of the formation fluid/gas using SCP location (between which two casings), its magnitude, their drilling and cement history, as well as their bleed-off and build-up trends. The pathway analysis is required for planning the requirements of the well intervention operation as well as assessment of risk involved. Li et al. (2018) shows an example analysis where the SCPs in the B and C annuli are respectively attributed to poor cementing of the 13 3/8-in casing and 7-in tie-back liner. As for SCP in the A-annulus, Zhang et al. (2017) suggested using Helium tracer method for leak detection (Fig. 4.13); this emerging technology and some gas expansion/contraction calculations help identify the depth of the leakage point of the completion casing. The main advantage of this method is the ability to test the well from the surface, and there is no need to shut in or choke the well. Fig. 4.14 presents the methodology of gas composition and isotope analysis for gas source detection. The best method for leak detection is using temperature noise logs (see Chapter 7) with several provider companies such as TGT, Archer, and Halliburton.

During well intervention operations, CBLs are run (see Chapter 4), temperature noise logs are run (see Chapter 7), etc. Sometimes, casing integrity is also checked using corrosion logs to measure the thickness of the casing pipe versus depth. Corrosion logs are not covered in detail in this book.

Ali et al. (2022) described a case study in an offshore well in the Persian Gulf by which, several weeks after cementing the 7-in liner, pressure buildup was noticed in three annuli of the casings. It is not clear if pressure buildup in those annuli had happened before or after running the 7-in liner. It is assumed that the pressure buildup was there but was undetected, or as a less possible alternative, maybe gas migration occurred during or after running/cementing the 7-in liner and gas penetrated in all three annuli of 30 × 24-in, 24 × 18 5/8-in, and 18 5/8 × 13 3/8-in casings with maximum SCP of 150 psi. Table 4.3 shows the maximum injection pressure of the resin system, which was determined for each case considering the burst resistance of the outer casing and the collapse resistance of the inner casing. It can be seen in the table that still greater maximum injection pressures could have been applied for the two outer annuli of 24 × 18 5/8-in, and 18 5/8 × 13 3/8-in. Following injection of a solids-free resin-based system in all the annuli, SCPs became zero and remained so for 1 year (at the time of reporting). As a matter of fact, the fluid has penetrated through the preexisting microchannels in the cement sheath to establish a dependable barrier against formation fluid's influx.

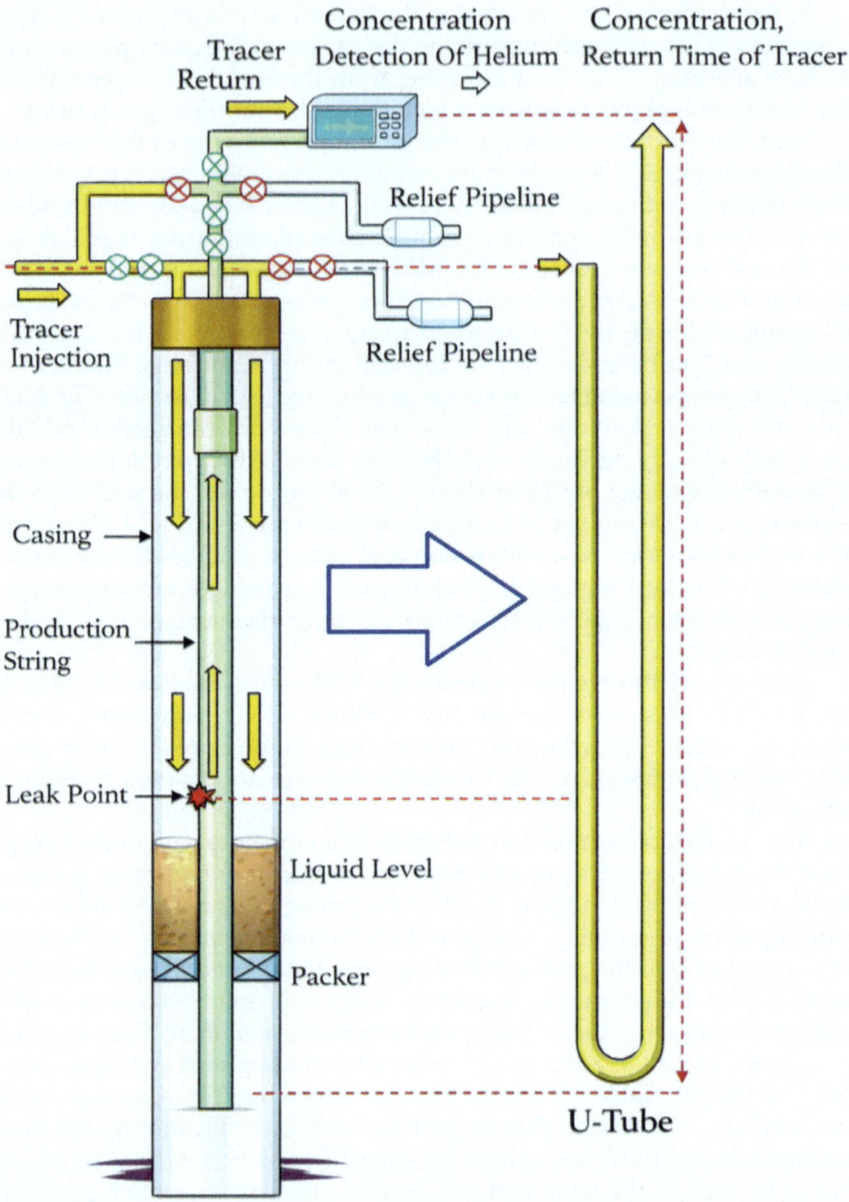

FIGURE 4.13 Leak detection in the completion string by helium tracer injection. *(Redrawn from Zhang, X., Fan, J., Liu, S., Wen, M., Lv, N., & Liang, Z. (2017). Diagnostic testing of gas wells with sustained casing pressure by application of He Tracer. ISOPE 2024. In Proceedings of the Twenty-seventh (2017). International Ocean and Polar Engineering Conference San Francisco, CA, USA June 25-30, 2017. https://publications.isope.org/proceedings/ISOPE/ISOPE%202017/data/64789-isope-vol4-1.3618547/t004-1.3619581/f009-1.3619619/a060-1.3619626/ap060-1.3619627.html).*

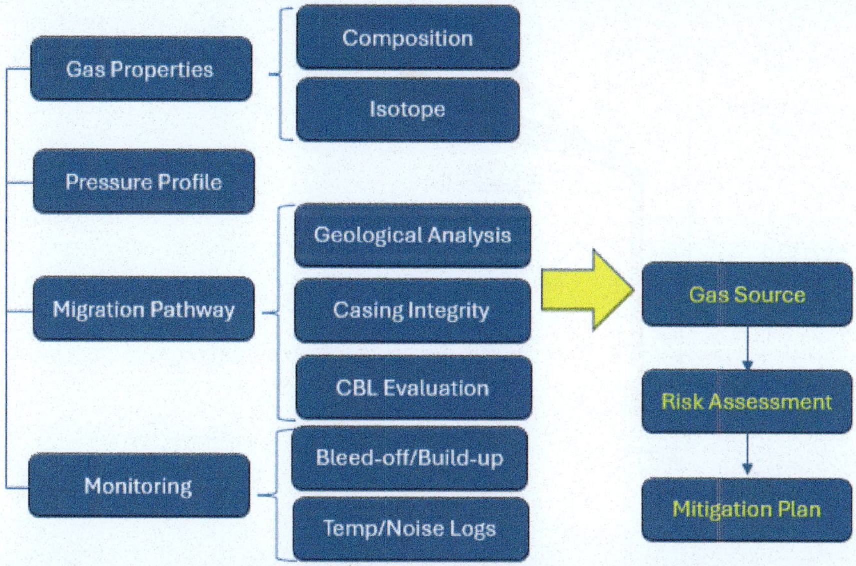

FIGURE 4.14 Gas source analysis for sustained casing pressure wells. *(Redrawn from Su, S., Vo, M., Jia, C., Yuan, J., Zhang, J., & Zhang, Q. (2021). Integrated gas source analysis for wells with sustained casing pressure: A case study from sour gas fields in Sichuan Basin, China. In Paper presented at the International Petroleum Technology Conference, Virtual. https://doi.org/10.2523/IPTC-21485-MS).*

TABLE 4.3 Annuli with sustained casing pressure (SCP) and determination of maximum injection pressure.

Annulus (CSG × CSG) (in)	SCP (psi)	Burst resistance of outer casing (psi)	Collapse resistance of inner casing (psi)	Max injection *P* (psi)
30 × 24	90	30-in: 1840	24-in: 776	**600** < 776
24 × 18 5/8	100	24-in: 2110	18-in: 1550	**900** < 1550
18 5/8 × 13 3/8	150	18 5/8: 3070	13 3/8: 1950	**900** < 1950

(Data from Kinik, K. & Wojtanowicz, A. K. (2011). Identifying environmental risk of sustained casing pressure. In Paper presented at the SPE Americas E&P Health, Safety, Security, and Environmental Conference, Houston, Texas, USA March 2011. https://doi.org/10.2118/143713-MS).

As another case study of treating with epoxy resin injection in Trinidad, SCP build-up trend was dramatically mitigated from 45 to 0.003 psi/hour after 1 year (Olsen et al., 2021).

4.7 Case study

An example of the consequence of ignoring monitoring SCP during production is presented by Ashena et al. (2011). The onshore oilfield is

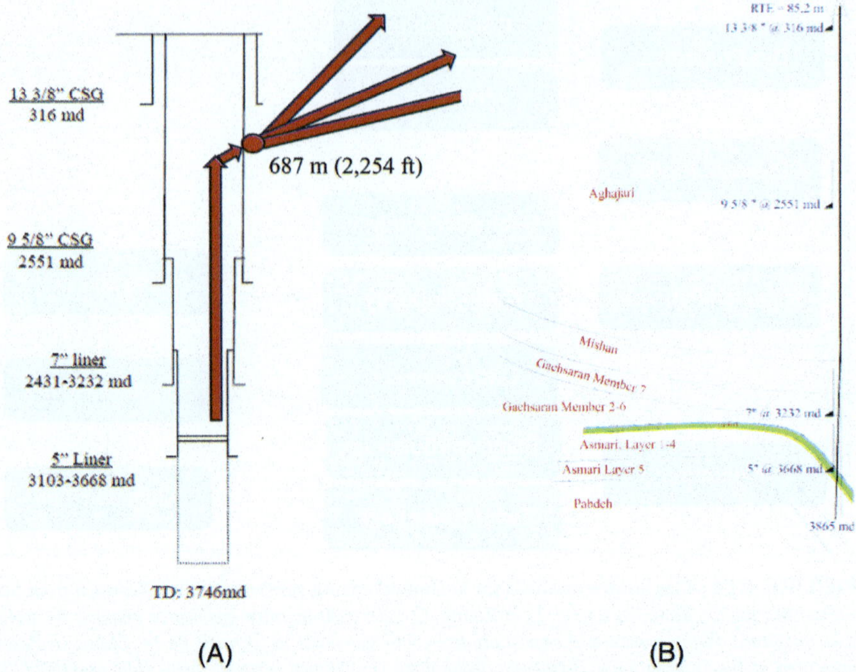

FIGURE 4.15 (A) Well construction of casings and liners and (B) well construction versus geological formations. *(Redrawn from Kumar, A., Gillies, J., Mehra, R., Pardeshi, M., Wilke, A., & Li, J. (2017). Improved primary cement jobs and mitigation of sustained casing pressure through the use of foamed spacer design technology. In Paper presented at the SPE/IADC Drilling Conference and Exhibition, The Hague, The Netherlands March 2017. https://doi.org/10.2118/184605-MS).*

located in the Middle East and is well known for its tectonically active features. In the wells drilled in this field, following running 18 5/8″ to the depth of 200–400 m (656–1312 ft), the intermediate 13 3/8″ casings were run to about 1200 m (3937 ft), and then the 9 5/8″ production casing was run. However, this incident well, which was drilled in 1976, had a different or nonstandard structure because the 18 5/8″ surface casing was eliminated, or say replaced by a lower size 13 3/8″ casing run to 316 m (1037 ft). Next, the production casing was run to 2551 m (8370 ft). The total depth of the well was 3746 m (12,290 ft). Fig. 4.15 shows the well construction and geological formations encountered by the well trajectory. During the 34 years of production, no proper measures for well integrity management were implemented by the well operator, and SCP was not monitored frequently. After this long term, the well probably suffered from a lack of well integrity in the casing and cement sheath.

Oil production rate had shown a decline for a while, and at least 50 seepages of fluid flow had appeared at the surface at the well site and all around the region. The well site was only 4 km away from the river supplying drinking water for three towns and the largest wildlife refuge of the country. Later, after controlling the blowout, it was identified that the 9 5/8″ casing and cement had undergone extensive degradation and erosion and finally leakage at the depth of 687 m (2254 ft); see Fig. 4.15. As the leak was located across the unconsolidated near-sandy and silty formations near the surface, gas had channeled/fractured the surface formation and percolated through the layers to the surface, spreading all around the region. The well was killed by a top-kill method, and then it underwent workover operation to go on stream/production.

References

Ali, W., Al-Turki, F. A., Abbas, A., Al-Yami, A. S., Wagle, V., & Dahmoush, A. A. (2022). Resin systems as evolving solution within the industry to replace the conventional remedial cementing while eliminating the sustained casing pressure SCP. In *Paper presented at the International Petroleum Technology Conference, Riyadh, Saudi Arabia* https://doi.org/10.2523/IPTC-21953-EA.

API. (2004). API technical report 17TR3, an evaluation of the risks and benefits of penetrations in subsea wellheads below the BOP stack. https://store.accuristech.com/searches/51662511

API. (2010). *API RP 65-2: Isolating potential flow zones in drilling and well completion operations* (2nd ed.). API.

API. (2006). *API RP-90: Annular casing pressure management for offshore wells*. API.

Ashena, R., Amani, M., Taei, B., & Kamyab, M. (2011). Management of the disastrous underground blowout in South of Iran. In *Paper presented at the SPE Middle East Oil and Gas Show and Conference, Manama, Bahrain* https://doi.org/10.2118/140781-MS.

Ashena, R., Thonhauser, G., & Dianati, D. (2014). Assessment of Reliability of Cement Bond Evaluation with Some Interesting Field Case Studies. Paper presented at the Abu Dhabi International Petroleum Exhibition and Conference, Abu Dhabi, UAE. https://doi.org/10.2118/172039-MS.

Ashena, R., Thonhauser, G., Vortisch, W., Rasouli, V., & Azizmohammadi, S. (2018). Optimization of core retrieval using a Thermo-Poro-Elastic (T-P-E) approach. *Journal of Petroleum Science and Engineering, 167*, 577–607. https://doi.org/10.1016/j.petrol.2018.04.041.

Ashena, R., Thonhauser, G., Ghalambor, A., Rasouli, V., & Manasipov, R. (2019). Determination of maximum allowable safe-core-retrieval rates. *SPE Reservoir Evaluation & Engineering, 22*(2019), 548–564. https://doi.org/10.2118/189480-PA.

Ashena, R., Bataee, M., Jafarpour, H., Abbasi, H., Zolotukhin, A., & Mirhashemi, M. (2021). Significant production improvement using optimization of completion and artificial lift: Case studies from South-West Iran. *Journal of Petroleum Exploration and Production Technology, 11*, 359–384. https://doi.org/10.1007/s13202-020-01053-6.

Ashena, R. (2025). *The practical handbook of well control*. Routledge, Taylor and Francis Group https://www.routledge.com/The-Practical-Handbook-of-Well-Control/Ashena/p/book/9781032753850?srsltid=AfmBOooZqOJyBoNQ8lWhWyMcNMmdX4-St1TUa2goxJzlN_6-31cO4lBo.

Bourgoyne, A. T., Scott, S. L., & Regg, J. B. (1999). Sustained casing pressure in offshore producing wells. In *Paper presented at the Offshore Technology Conference, Houston, Texas* May 1999 https://doi.org/10.4043/11029-MS .

Brodie, A. (2016). The autonomous management of wells with unplanned sustained casing pressure. In *Paper presented at the Abu Dhabi International Petroleum Exhibition & Conference, Abu Dhabi, UAE* November 2016 https://doi.org/10.2118/183077-MS .

Burgoyne, A. T., Scott, S. L., & Manowski, W. (1997). *A review of sustained casing pressure. Outer continental shelf's wells*. Bureau of Safety and Environmental Enforcement (BSEE).

Clark, B. E., Pearl, W., & Hopwood, D. (2023). Mitigating sustained casing pressure in the Denver-Julesburg basin with a low permeability flow resistant cementing solution. In *Paper presented at the International Petroleum Technology Conference, Bangkok, Thailand* https://doi.org/10.2523/IPTC-22872-EA.

Ezani, F. S., Thuzar, M., Kumar, A. K., & Lau, C. H. (2021). Prevention of historically challenging sustained casing pressure from shallow gas through re-engineered cement design and execution methodology. In *Paper presented at the International Petroleum Technology Conference, Virtual* March 2021 https://doi.org/10.2523/IPTC-21429-MS .

Ghalambor, A., Mikolajczy, R., & Revett, L. W. (2000). A prelude to formation damage prevention-achieving a reliable primary cement job. *SPE paper no. 58770, Proceedings of the 14th International Symposium on Formation Damage Control* (pp. 537–546). Louisiana: Lafayette 23-24 February 2000.

Holand, P. (2017). Loss of well control occurrence and size estimators, Phase I and II. Report No. ES201471/2. Office/Division Program. TAP. Project Number. 765. Category. Deepwater.

Huerta, N. J., Checkai, D. A., & Bryant, S. L. (2009). Utilizing sustained casing pressure analog to provide parameters to study CO_2 leakage rates along a wellbore. In *Paper presented at the SPE International Conference on CO_2 Capture, Storage, and Utilization, San Diego, California, USA* November 2009 https://doi.org/10.2118/126700-MS .

ISO (2017). ISO 16530-1: International standard, petroleum and natural gas industries—Well integrity—Part 1: Life cycle governance (1st ed.). https://www.iso.org/standard/63192.html.

Kinik, K., & Wojtanowicz, A. K. (2011). Identifying environmental risk of sustained casing pressure. In *Paper presented at the SPE Americas E&P Health, Safety, Security, and Environmental Conference, Houston, Texas, USA* March 2011 https://doi.org/10.2118/143713-MS.

Langley, M., Cleveland, M., Eulberg, J., & Hudson, M. (2018). Flexible, self-healing cement eliminates sustained casing pressure in Denver-Julesburg Basin unconventional wells. In *Paper presented at the IADC/SPE Drilling Conference and Exhibition, Fort Worth, Texas, USA* March 2018 https://doi.org/10.2118/189587-MS .

Kumar, A., Gillies, J., Mehra, R., Pardeshi, M., Wilke, A., & Li, J. (2017). Improved primary cement jobs and mitigation of sustained casing pressure through the use of foamed spacer design technology. In *Paper presented at the SPE/IADC Drilling Conference and Exhibition, The Hague, The Netherlands* March 2017 https://doi.org/10.2118/184605-MS .

Landry, G., Welty, R. D, Thomas, M., Vaughan, M. L., & Tatum, D. (2015). Bridging the gap: An integrated approach to solving sustained casing pressure in the Cana Woodford shale. In *Paper presented at the SPE Well Integrity Symposium, Galveston, Texas, USA* June 2015 https://doi.org/10.2118/174525-MS .

Li, X., Wang, C., Wang, Y., Li, F., & Xu, H. (2018). Practice of sustained casing pressure removal for JZ gasfield production well plug and abandonment in Bohai Bay. In *Paper presented at the SPE Symposium: Decommissioning and Abandonment, Kuala Lumpur, Malaysia* December 2018 https://doi.org/10.2118/193944-MS .

McDonald, M., & Li, X. (2017). The development & field results of a new, advanced form of sodium silicate as a cost effective solution for treatment for sustained casing pressure. In *Paper presented at the SPE Thermal Well Integrity and Design Symposium, Banff, Alberta, Canada* November 2017 https://doi.org/10.2118/188166-MS .

MMS. (2010). *30 CFR Part 250: Annular casing pressure management for offshore wells*. Department of Interior, Mineral Management Service.

Nabaei, M., Moazzeni, A., Ashena, R., & Roohi, A. (2011). Complete loss, blowout and explosion of shallow gas, infelicitous horoscope in middle east. In *Paper presented at the SPE European Health, Safety and Environmental Conference in Oil and Gas Exploration and Production, Vienna, Austria* February 2011 https://doi.org/10.2118/139948-MS .

Nafikova, S., Bugrayev, A., Taoutaou, S., Baygeldiyev, G., Akhmetzianov, I., Gurbanov, G., & Eliwa, I. (2019). Elimination of the sustained casing pressure using self-healing cement in Turkmenistan section of the Caspian Sea. In *Paper presented at the SPE Annual Technical Conference and Exhibition, Calgary, Alberta, Canada* September 2019 https://doi.org/10.2118/195945-MS .

Nafikova, S., Ramazanova, Y., Muslimov, A., Akhmetzianov, I., Jain, B., Kim, A., & Zvyagin, V. (2021). Lessons learned and case studies of overcoming sustained casing pressure in extended-reach wells. In *Paper presented at the SPE Western Regional Meeting, Virtual* April 2021 https://doi.org/10.2118/200792-MS .

Olsen, J. E., Hosein, W., Ringe, T., & Friedli, J. D. (2021). Downhole annular barrier solution for sustained casing pressure – Trinidad case study. In *Paper presented at the SPE Trinidad and Tobago Section Energy Resources Conference, Virtual* June 2021 https://doi.org/10.2118/200976-MS .

Reddy, B. R., Xu, Y., Ravi, K., Gray, D., & Pattillo, P. D. (2009). Cement-shrinkage measurement in oilwell cementing—A comparative study of laboratory methods and procedures. *SPE Drilling & Completion, 24*(2009), 104–114. https://doi.org/10.2118/103610-PA.

Rocha-Valadez, T., Hasan, A.R., Mannan, M.S., & Kabir, S. (2014). Assessing wellbore integrity in sustained-casing-pressure annulus, SPE Drilling & Completion, 29(01), 131–138. Doi:10.2118/169814-PA.

Su, S., Vo, M., Jia, C., Yuan, J., Zhang, J., & Zhang, Q. (2021). Integrated gas source analysis for wells with sustained casing pressure: A case study from sour gas fields in Sichuan Basin, China. In *Paper presented at the International Petroleum Technology Conference, Virtual* https://doi.org/10.2523/IPTC-21485-MS.

Wang, Y., Gao, D., & Wang, J. (2020). New model for prediction of sustained casing pressure of deepwater wells. In *Paper presented at the 30th International Ocean and Polar Engineering Conference, Virtual* October 2020.

Wang, Y., Xin, S., Gao, D., Wang, J., Ning, B., & Zhao, Z. (2022). Sustained casing pressure: One of the most important factors determining the integrity of deepwater wells. In *Paper presented at the Offshore Technology Conference, Houston, Texas, USA* May 2022 https://doi.org/10.4043/31759-MS .

Zeng, J., Gao, D., Wang, Y., Fang, J., & Liu, K. (2018). Research on the effect of casing deformation on sustained casing pressure. In *Paper presented at the 52nd U.S. Rock Mechanics/Geomechanics Symposium, Seattle, Washington.*

Zhang, X., Fan, J., Liu, S., Wen, M., Lv, N., & Liang, Z. (2017). Diagnostic testing of gas wells with sustained casing pressure by application of He Tracer. ISOPE 2024. In *Proceedings of the Twenty-seventh (2017) International Ocean and Polar Engineering Conference San Francisco, CA, USA* June 25-30, 2017 https://publications.isope.org/proceedings/ISOPE/ISOPE%202017/data/64789-isope-vol4-1.3618547/t004-1.3619581/f009-1.3619619/a060-1.3619626/ap060-1.3619627.html .

Zhang, B., Guan, Z., Lu, N., Hasan, A. R., Wang, Q., & Xu, B. (2019). Trapped annular pressure caused by thermal expansion in oil and gas wells: A review of prediction approaches, risk assessment and mitigation strategies. *Journal of Petroleum Science and Engineering, 172*, 70–82. https://doi.org/10.1016/j.petrol.2018.09.041.

CHAPTER

5

Cement bond logs and evaluation

Rahman Ashena, William K. (Bill) Ott

Smart Energy Data Analytics (SEDA)-Group, Houston, TX, United States

5.1 Introduction

This chapter presents a comprehensive overview of cement bond logs (CBLs), covering both the fundamental principles and practical considerations involved in their use. It explores key wellbore conditions that influence CBL readings, outlines quality control and evaluation guidelines, and includes field case studies that illustrate real-world challenges and solutions in ensuring well integrity.

CBLs are essential diagnostic tools in the oil, gas, and geothermal industries, used to evaluate the effectiveness of a cement job by measuring the bond between the casing and the surrounding formation. By detecting zones where cement has failed to provide adequate isolation, CBLs help prevent fluid migration and support long-term wellbore integrity. These logs are typically acquired using acoustic tools that assess bond quality and identify potential voids or channels within the cement sheath. One contributing factor to the 2010 Gulf of Mexico Deepwater Horizon blowout was the absence of CBLs (Ashena, 2025).

The evolution of CBL technology dates back to the early 20th century, with acoustic logging tools introduced in the 1960s. Over the past seven decades, technological advancements have significantly enhanced the accuracy, resolution, and diagnostic capabilities of cement bond evaluation. Although early tools offered basic insights into casing-cement bonding, modern systems can complement traditional sensors to provide high-precision data that are integral to comprehensive well integrity programs.

Well Cement Integrity
DOI: https://doi.org/10.1016/B978-0-443-51564-4.00016-3

Bridging theory and practice, this chapter equips readers with both the technical foundation and applied understanding needed to effectively interpret CBL data and make informed decisions in cement evaluation and remediation.

5.2 Conventional cement bond log

Conventional cement bond logging has been in use since the 1960s. It employs a piezoelectric transmitter that emits low-frequency acoustic waves (18–20 kHz) in bursts of approximately 50 microseconds, with repetition rates between 10 and 60 Hz. These waves propagate through the casing and its surrounding materials, enabling measurement of the bond quality between casing and cement.

The logging tool includes a transmitter and two piezoelectric receivers positioned at 3 and 5 ft from the transmitter. Older versions had the second receiver at 4 ft. Centralizers are used to prevent tool eccentricity. Updated versions, like radial CBL tools, can include up to eight receivers for azimuthal cement evaluation.

Acoustic waves refract into the casing wall, generating a casing wave when entering at the critical angle. This casing wave is essential for evaluating the bond between the casing and its surroundings.

5.2.1 Deliverables

There are several deliverables by the conventional CBL tool: the transit time (TT), the amplitudes of the first and subsequent waves received, and the attenuation rate (ATT). There are geological gamma ray and casing collar log that are just used for geological formations and depth matching. The deliverables are displayed in different forms depicted in the following sections.

5.2.2 Cement bond log value

The CBL value is the peak amplitude of the first wave received at the 3-ft receiver. The amplitude is affected by shear coupling between casing and annulus. Well-cemented casing leads to greater attenuation and lower CBL values. Free-pipe conditions, where little or no cement is present, result in high CBL values (often >50 mV), whereas good cement bonds show lower values (<20 mV).

5.2.3 Composite wave

Composite waves are received signals made up of casing, shear, longitudinal, and mud waves. Their components vary with the annular fill and formation type. Variable density log (VDL) displays are used to visualize these waveforms across depth.

There are two display methods:

✓ Waveform Display: Discrete depth intervals with full waveform plots.
✓ *VDL*: Continuous display of wave amplitudes using grayscale, where dark shades represent higher amplitude.

Casing waves appear as straight lines on the left side of VDL. Formation waves, visible on the right, are wavy and indistinct if well-cemented but clear and straight if bonding is poor.

The VDL uses the 5-ft receiver for better resolution. Interpretation remains qualitative but is crucial for evaluating cement bond quality alongside CBL values and TT. Fig. 5.1 shows a typical example of good and bad bond example using typical CBL and VDL displays.

5.2.4 Attenuation rate and bond index

ATT quantifies energy loss between receivers by relating the amplitudes of the first and second half cycles of the first wave received.

Bond index (BI) links attenuation to cement coverage. $BI = 1$ indicates 100% cement coverage.

BI is displayed alongside CBL for hydraulic isolation assessment. Industry standards recommend $BI \geq 0.8$ for good cement in hard formations and ≥ 0.6 in unconsolidated zones. However, BI alone may misrepresent bonding, especially in cases of tool eccentricity or varying compressive strengths.

Hydraulic isolation requires a minimum interval of effective bonding. This interval varies with casing diameter and BI. For example, a 7-in casing with $BI = 0.8$ needs 11 ft of bonded length, whereas $BI = 0.6$ needs 19 ft. BI-based cement bond and hydraulic isolation assessments have limitations due to several factors:

✓ BI depends solely on CBL values, which can be unreliable in cases like tool eccentricity.
✓ CBL/BI mainly reflects cement at the casing interface, not the formation interface.
✓ Both annular fill and compressive strength affect CBL/BI, but downhole data on these is often unknown.

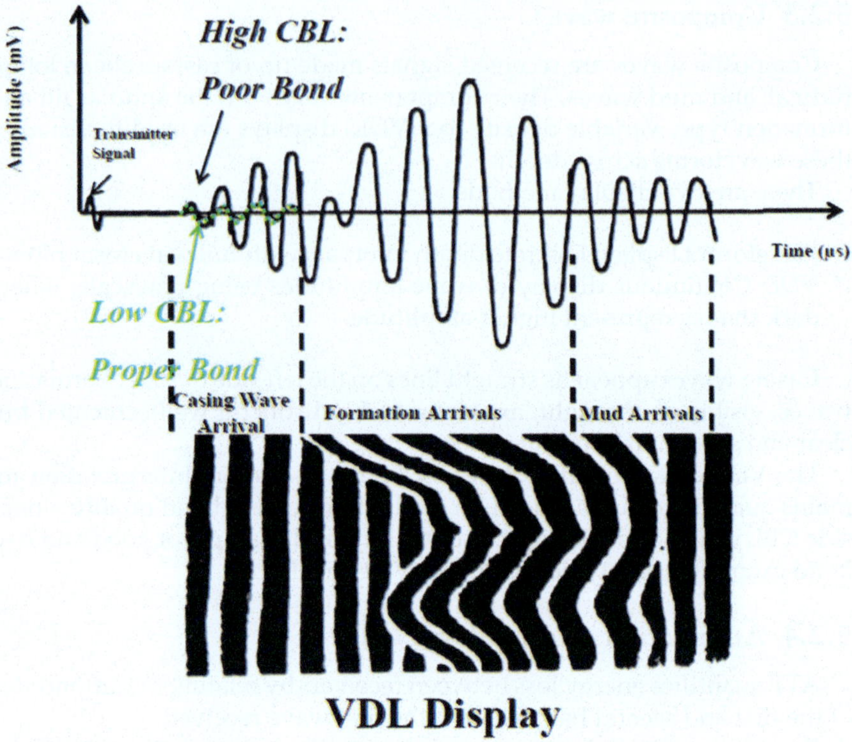

FIGURE 5.1 Good and poor bonds deduced by considering cement bond log (CBL) values and variable density log (VDL) displays. *(Redrawn from Smolen, J. (1996). Chapter 11, Well integrity: Ultrasonic pulse echo cement evaluation, and Chapter 10, Well integrity: Acoustic cement evaluation survey. Cased hole and production log evaluation. PennWell).*

As a result, CBL and BI alone cannot confirm hydraulic isolation. For a more accurate evaluation, incorporate TT, VDL, ultrasonic logs, and other supporting methods in the holistic approach. Remedial cementing should be reserved for confirmed free-pipe cases, with full understanding of the risks.

5.3 Updated technologies

5.3.1 Radial cement bond log

Radial CBL tools enhance conventional CBL by including multiple receivers arranged circumferentially. Tools, such as Halliburton's RCBS and RCBL, have six and eight receivers, respectively. These tools generate cement amplitude maps, where dark colors indicate good bonding and light colors indicate poor bonding.

5.3.2 Electromagnetic technology

Electromagnetic acoustic transducer (EMAT)-based tools, like Baker Hughes' segmented bond tool, use pads in direct contact with the casing wall and generate both compressional and shear waves. Benefits include:

✓ Independence from wellbore fluid
✓ Microannulus identification without pressurization
✓ Reliable top-of-cement detection
✓ Shear and flexural attenuation mapping is provided
✓ Operation in gas-filled wells
✓ No need for logging tools' centralizers (pads contact casing directly)

EMAT tools provide detailed radial cement maps and dual-attenuation plots, helping to discriminate between poor bonding and fluid-filled gaps.

5.3.3 Ultrasonic cement bond log

Ultrasonic tools operate using pulsed echo technology at high frequencies (200–700 kHz). These tools offer:

✓ Higher azimuthal resolution
✓ Direct mapping of casing and cement properties
✓ Enhanced channel detection

They emit high-frequency waves that resonate the casing and return via the borehole fluid. Decay rates indicate the acoustic impedance behind the casing. Faster decay implies solid (cement) presence, whereas slow decay indicates fluids.

Ultrasonic tools are highly effective when used alongside sonic tools. Although radial tools map the full circumference, ultrasonic tools more precisely identify void locations. They are less affected by mud weight, tool eccentricity, fast formations, and human errors.

5.4 Evaluation

This section covers cement bond evaluation, beginning with a quality control procedure, followed by the evaluation methodology, and concluding with a discussion of factors that can influence cement bond readings.

5.4.1 QC

CBLs are not necessarily always reliable. To ensure quality control of CBLs and the validity of interpretations, several factors must be considered. Failure to do so may render the log invalid. Common issues include

TABLE 5.1 QC features and what is required to do.

	QC feature	What to do
1.	Calibration	Ensure logging tool was properly calibrated
2.	Tool prerun Check	Verify tool functionality and secure accessories
3.	Surface Pressure	Apply surface pressure to eliminate microannulus effect
4.	Multiple Pipes	Use special ultrasonic CBL for double/multiple casings
5.	Eccentricity	Confirm centralizers were used during casing pipe running and logging
6.	Transit Time	Varying excessively = eccentricity; too flat = likely fake or unrepresentative
7.	Repeat section	Must match main log for reliability

CBL, cement bond log.

lack of tool calibration, eccentricity, microannulus effect, and mistakes or manipulation by the logging crew. Table 5.1 and Fig. 5.2 illustrate seven criteria for QC of logs.

As the TT and repeat-section curves are critical for verifying CBL reliability, they are further explained in this section.

5.4.1.1 *Transit time curve*

TT is a key indicator of log reliability. It should always accompany the CBL value and VDL display. If the TT graph is unavailable alongside the CBL/VDL for any reason, it becomes impossible to perform effective quality control and ensure the reliability of the CBLs.

TT is the time between wave transmission and reception of the first wave at the 3-ft receiver in a typical CBL tool. TT depends on casing dimensions and acoustic properties in the mud and casing pipe. Table 5.2 shows deviations from predicted TT values and the corresponding indications.

5.4.1.2 *Repeat-section curve*

Repeat logging enhances reliability. Table 5.3 lists the repeat-section intervals depending on cases. Note that borehole and logging conditions must remain constant to validate comparisons between main and repeat logs.

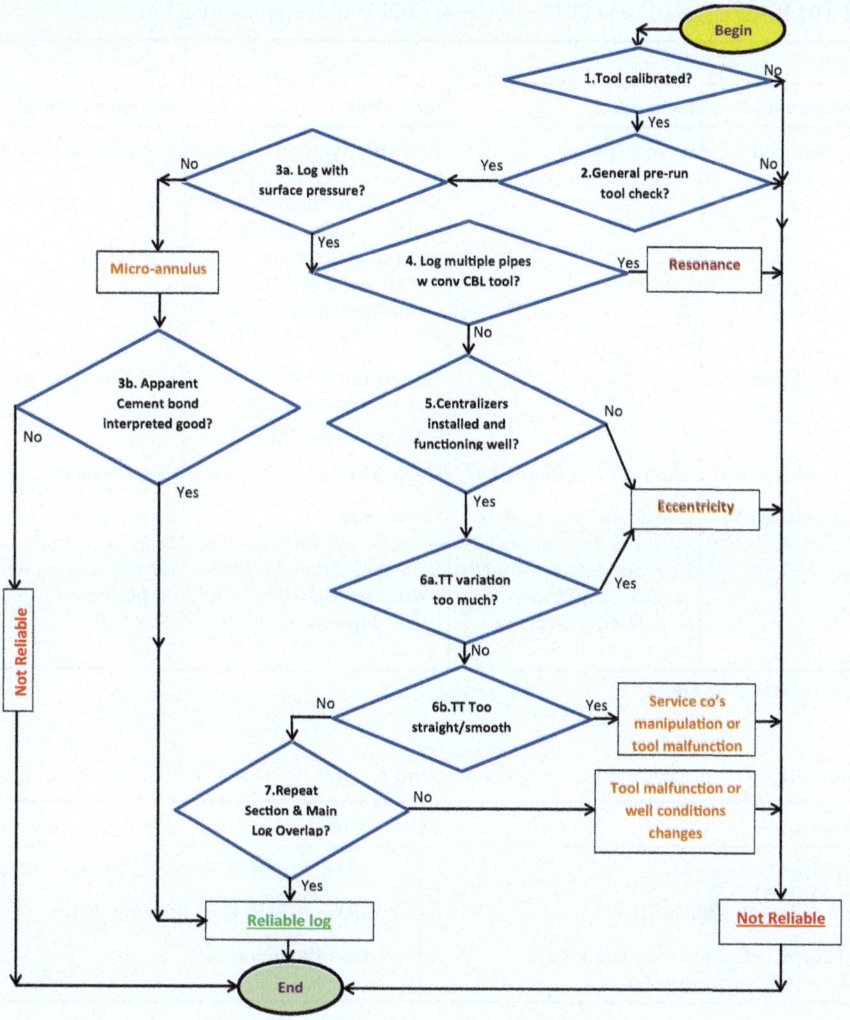

FIGURE 5.2 A flowchart for QC of cement bond logs. *TT*, Transit time (Ashena and Ott, 2025).

5.4.2 Evaluation

Cement bond evaluation typically involves both conventional CBL tools—including radial CBL tools—and ultrasonic CBL tools. As ultrasonic CBL is often conducted alongside conventional CBL, a comprehensive evaluation must consider parameters from both methods. For conventional CBL, the evaluation includes parameters such as ATT, CBL amplitude, BI, VDL display, and cement maps (for radial tools). For ultrasonic CBL, key

TABLE 5.2 Transit time (TT) cases, indications, and interpretation hints.

TT			
Observation	**How much**	**Indication**	**Interpretation hint**
<Predicted TT	up to 70 μs	**Eccentricity** of the logging tool or casing/liner pipe *Note*: The segmented bond tool uses contact pads during deployment, so it does not require centralizers and is less affected by eccentricity	Not reliable (except for SBT)
<Predicted TT	Up to 4–5 μs	**Fast formations** *Note*: double check the geology using well logs	Excellent cement
>Predicted TT	Minimal: less than 15 μs	**Cycle-stretch**	Good cement
>Predicted TT	Considerable: 15–50 μs *Note*: The first half-cycle E1 is not detected due to too low amplitude (< detection level of logging tool)	**Cycle-skip** *Note*: Recent compensated tools with more receivers reduce the risk of cycle-skipping	Excellent cement *Note*: bond to the formation may not be perfect

SBT, *segmented bond tool.*

TABLE 5.3 Repeat sections for cement bond logging.

When	Where
Always	100 m (330 ft) from casing/liner shoe up
Liner length >500 m (1640 ft)	100 m (330 ft) from liner lap down
If suspicious data or anomalies are detected during logging.	Additional interval

parameters include ATT, acoustic impedance, and cement maps. Table 5.4 summarizes the evaluation criteria used for both good and poor cement bond conditions across these tools. All listed parameters should be reviewed collectively to ensure accurate assessment of cement bond integrity.

5.4.2.1 Factors

Interpreting cement bond logs (CBL) can be complex, as the measurements are influenced not only by the cement sheath and coverage but also by various other factors, including wellbore conditions. This section explores these influencing factors and their impacts. Several key properties

TABLE 5.4 Overall evaluation for sonic and ultrasonic cement bond logs (CBLs).

Parameter		Cement bond quality	
		Good	Poor
Conventional CBL			
Attenuation rate		**High**	**Low**
CBL value		**Low (<20 mV)**	**High (>50 mV)**
Bond index		**High (>0.6–0.8)**	**Low (<0.6)**
VDL		**Wavy indistinct lines**	**Vertical distinct lines**
Cement map (radial CBL)		**Dark colors**	**Light colors**
Ultrasonic CBL			
Attenuation rate	(For each receiver) Average	**High**	**Low**
Acoustic impedance	(For each receiver) Average	**High**	**Low**
Cement map	Processed	**Dark**	**Light**

VDL, *variable density log.*

are examined, with comprehensive discussions provided by Ashena and Ott (2025); Table 5.5 presents a summary of their findings.

5.5 Case studies

This section aims to familiarize readers with the interpretation skills required for analyzing CBLs. To achieve this, two sonic and ultrasonic CBL field case studies are presented and interpreted. At the bottom of each log the evaluation of the cement bond is qualitatively provided as good, intermediate, or poor. The second case study in this chapter is briefing of a modification of a case presented by Ashena and Ott (2025) and Ashena et al. (2014).

5.5.1 Case study I: Conventional cement bond log across casing

Fig. 5.3 shows the CBL/VDL logs across the 5.5-in casing from 4410 to 4600 ft. Interpretation of the logs is presented with the following points:

✓ Please note that the CBL/VDL logs do not cover the double-pipe intervals; therefore this does not compromise the reliability of the logs.

TABLE 5.5 Effects of some factors affecting cement bond logs.

Factor	Detection methods	Effect on log and interpretation
Annular fill	Review cementing operation/records and check with software or lab simulation	Annular fill↑ → annular sealing and wave attenuation↑ Better bond is observed in log. E.g., in conventional CBL: CBL value↓ or BI↑
Cement properties	Review lab and cementing operation	Cement strength↑ → wave attenuation ↑ (Controversial parameter and its effect may be mixed with annular fill) It has a positive display on observation (though compressive strength is not, by itself, enough) Assuming the same annular fill, bond quality↑. In conventional CBL, Attenuation-rate↑, CBL value↑ or BI↓, or acoustic impedance↓
Parameters adjustment	Monitor logging crew and their activities during and after logging	1. The combination of two parameters, "fixed-gate" and "detection level," in conventional CBL logging can contribute to cycle-skip occurrence 2. Logging crew's usually unintentional mistake (not possible in all cases to determine appropriate values) 3. Transit time smoothing by crew (intentional manipulation) 4. It can cause unreliable/unrepresentative log and interpretation
Pressure and temperature	• Review formation pressure and temperature from other logs • And check well operations or fluid injected or produced before cement bond logging	1. High temperatures in deep or geothermal wells can alter the mechanical properties of the cement, including its acoustic impedance 2. High pressures or temperatures cause creation of microannulus or microchannel 3. CBL tools may not operate well at too high wellbore temperatures *Note*: In severe cases, it may cause the log not to be fully representative
Formation type	• Use open-hole logs • Check transit time curve (a slight drop in TT may indicate fast formations)	1. Shale has sloughing characteristics contributing to bond. Better bond is usually across shale beds 2. Fast formations mean well-cemented carbonate rocks to casing and formation. Across fast formations, though greater CBL values than real values are observed, the bond should be interpreted as excellent

(*continued on next page*)

TABLE 5.5 Effects of some factors affecting cement bond logs—cont'd

Factor	Detection methods	Effect on log and interpretation
Mud properties and mud cake	Review mud reports and indications during cementing and displacement	1. a. Mud weight↑ → Attenuation of waves↑ b. Next, many CBL tools, even ultrasonic, cannot work well at very high mud weights above 16 ppg. The bond quality interpreted from a CBL may have some discrepancy from the real/representative bond 1. The presence of thick mud cake against the formation wall with water-based muds may lead to channels between the cement sheath and the formation. Poor bond from logs may be observed in case of thick mud cakes
Pipe diameter	Review casing pipe size	Different wave attenuation occurs in different pipe diameters. ✓ For the free-pipe state, Pipe diameter↑ → CBL value↓ (significantly) ✓ For the well-cemented state, Pipe diameter↑ → CBL value↑ (slightly) Thus the range of CBL value is significantly less for larger diameter pipes
Multiple pipes	Review well schematic to see casing and liner pipes overlap	There would be resonance and interference of the readings of acoustic waves from the annuli of the next outer casing pipes The log would be mostly unreliable and unrepresentative unless special logging tools are used. With conventional tools, it usually shows as apparently poor cement bond, but the reality may be different
Eccentricity	• Review casings running reports (if pipes centralizers were not used) and if there was presence of excessive scale on the inner wall of the pipe • Witness and review cement bond logging (if logging tool centralizers were not used)	✓ Eccentricity causes interference of signals returning to receiver from all directions ✓ Eccentricity typically manifests itself as a reduction in both TT and CBL values ✓ It results in unreliable and unrepresentative log and interpretation

(continued on next page)

TABLE 5.5 Effects of some factors affecting cement bond logs—cont'd

Factor	Detection methods	Effect on log and interpretation
Microannulus and microchannel	• Review well operations • Too much lubricant/dope used on casing	1. Microannulus does not make the cement bond poor but would most likely cause log misleadingly show poor bond. Unreliable/unrepresentative log and interpretation for microannulus if logging is not done underpressure 2. Microchannel does definitely make the bond quality and log poor and representative. Reliable log when microchannels are there
Underpressure logging	Review CBL logging operations	Eliminates the microannulus effect, but it cannot eliminate microchannels. It makes the log representative

BI, *bond index;* CBL, *cement bond log;* TT, *transit time.*

TABLE 5.6 Cement bond evaluation of the ultrasonic USIT log (case study II).

Interval (m)	USI interpretation	Evaluation of cement quality
3566–3575	Existence of blue color in track 1, considerable liquid percentage behind pipe (blue color in track 2), and intermediate values for average acoustic impedance (track 3)	Lower intermediate
3575–3600	Existence of black color in track 1, considerable cement percentage behind pipe (yellow color in track 2), and high value for average acoustic impedance (track 3)	Good

✓ The TT curve shows only slight fluctuations, which suggests that the logging tool was nearly centered during the logging operations. In addition, as the curve is not completely smooth (varies slightly), it indicates that there was likely no log manipulation by the logging crew.

✓ At about 4565 ft a cycle skip is observed, indicated by an abnormally large TT value. This issue appears to have been recognized by the tool and/or the logging crew, as reflected by the corresponding low CBL value.

✓ The cement quality is very good with low CBL values, wavy indistinct VDL, and dark cement maps. Just at 4420 ft the cement quality is intermediate with CBL value of around 30 mV, and the VDL is less wavy, and some light colors are visible in the cement map.

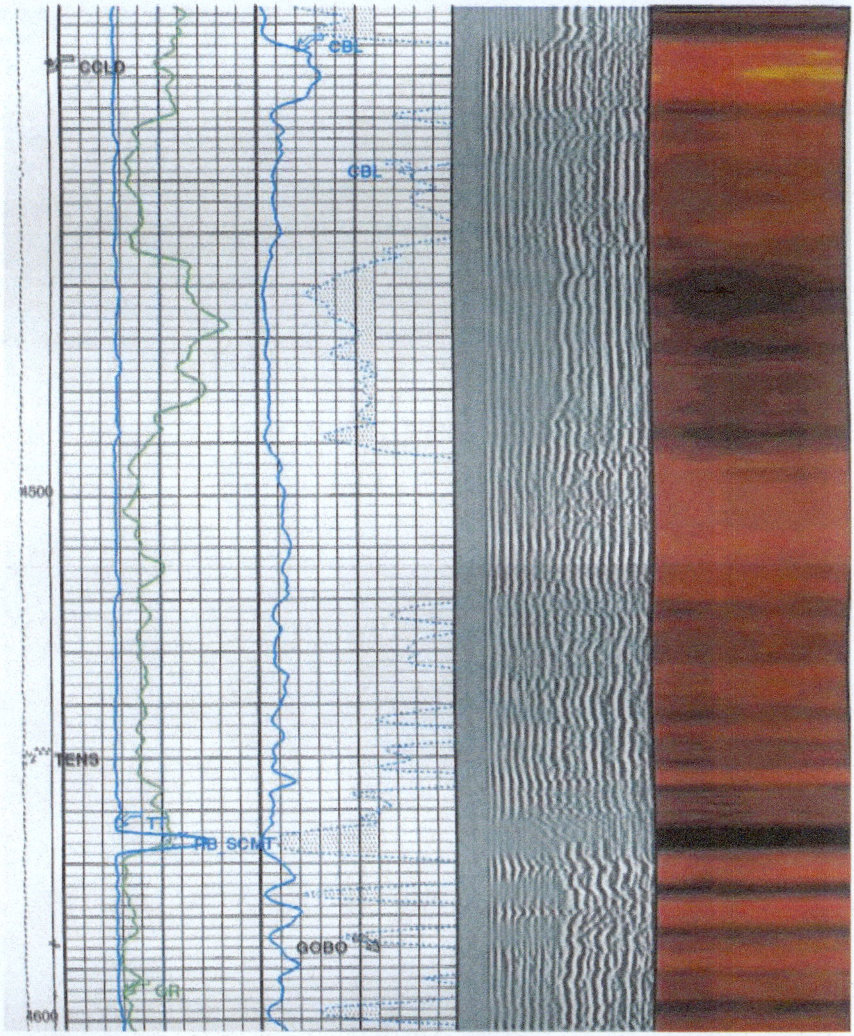

FIGURE 5.3 Case study I: Conventional cement bond log (CBL) in a field in the United States (Courtesy Breakthrough Ventures LCC).

5.5.2 Case study II: USIT log along with conventional cement bond log

Table 5.6 and Fig. 5.4 show the SLB's ultrasonic USIT log over a 51-m (168 ft) interval (3549–3600 m/11,644–11,812 ft). Key USI log tracks (right to left) include:

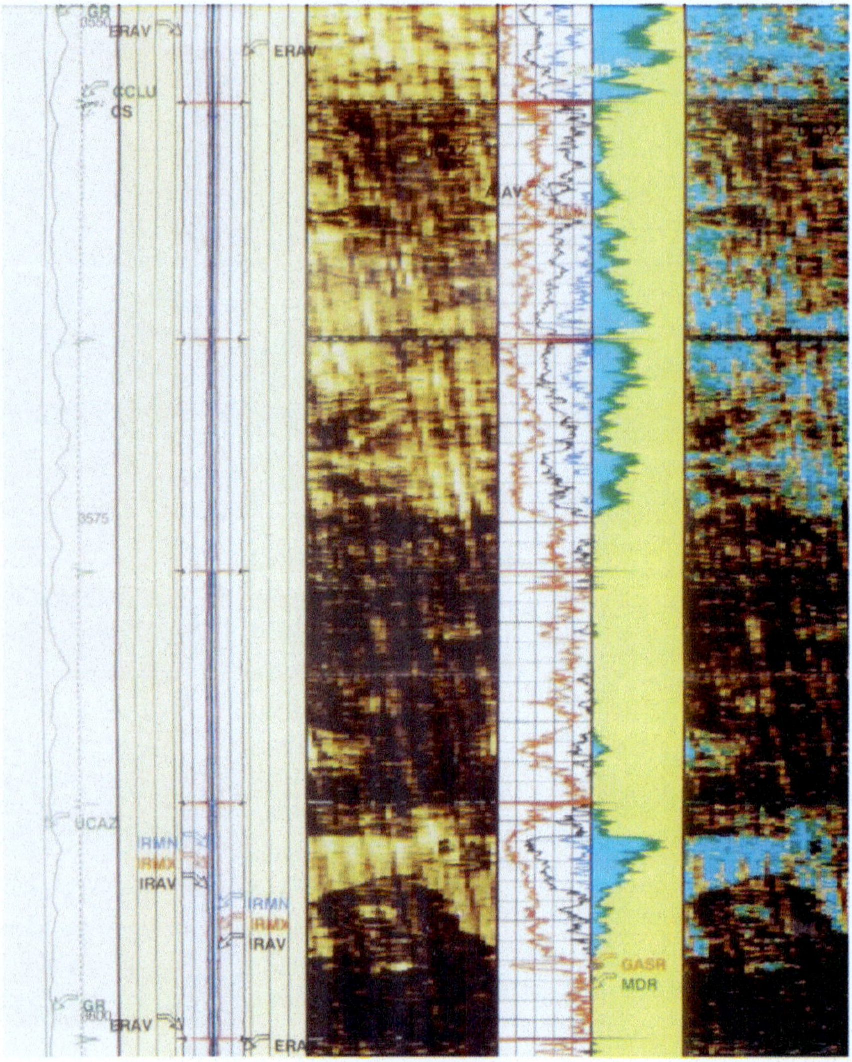

FIGURE 5.4 Case study II: The USI log across a 7-in liner of a well from one of the Middle Eastern Oil Field for the interval 3549–3600 m (11,644–11,812 ft).

✓ Track 1: Azimuthal acoustic impedance map (0–360 degrees); darker shades indicate better cement bond.

✓ Track 2: Volumetric cement map—yellow (cement), blue (liquid), red (gas); darker yellow indicates better bond; higher blue suggests possible hydraulic communication. This track also serves as a BI, increasing from right to left.

- ✓ Track 3: Acoustic impedance graphs (min, avg, and max); values range from 0 to 4 MRayls, with rightward trends showing stronger bonding.
- ✓ Track 4: Raw acoustic impedance data, not used for interpretation.

A summary of CBL evaluation is provided in Table 5.4.

References

Ashena, R., & Ott, W. K. (2025). *Cement bond logs guidelines and evaluation*. Springer Publishing.

Ashena, R., Thonhauser, G., & Dianati, D. (2014). Assessment of reliability of cement bond evaluation with some interesting field case studies. In *Paper presented at the Abu Dhabi International Petroleum Exhibition and Conference, Abu Dhabi, UAE* https://doi.org/10.2118/172039-MS.

Ashena, R., Bahreini, H., & Lotfi, M. (2023). Statistical analysis of past kicks and blowouts occurred in a Middle Eastern oilfield. *Journal of Petroleum Exploration and Production Technology, 13*, 2063–2082. https://doi.org/10.1007/s13202-023-01664-9.

Ashena, R. (2025). *The practical handbook of well control* (1st ed.). CRC Press https://doi.org/10.1201/9781003473770.

Smolen, J. (1996). Chapter 11, Well integrity: Ultrasonic pulse echo cement evaluation, and Chapter 10, Well integrity: Acoustic cement evaluation survey. *Cased hole and production log evaluation*. PennWell.

CHAPTER

6

Pressure integrity tests

Rahman Ashena

Smart Energy Data Analytics (SEDA)-Group, Houston, TX, United States

Reviewed by Mr. Stan Christman, Independent Consultant

6.1 Introduction

Well, barriers should undergo pressure testing to confirm their integrity and ensure safe operation. Accordingly, an accurate evaluation of the cement and formation integrity is extremely important during drilling and workover operations in a well. A statistical work by Holand (2017) shows that poor cement was cause of 25% of loss of well control events during deep-zone drilling using floaters in the US Gulf of Mexico (GOM). Another study by Ashena et al. (2023) in the Middle East (ME) shows that 6 out of all 117 kicks (5.13%) in an oilfield's history occurred during cement drill-out, which indicates bad/nonintegrated shoe-track cement.

A hydraulic pressure integrity test (PIT) is usually performed after the cementing job to evaluate the isolation provided by the cement. These tests are either performed before casing shoe drill-out or after that. There are two types of tests carried out to check the integrity of well barriers: the positive-pressure test (PPT) and negative-pressure test (NPT), or inflow test. In a PPT, well barriers are usually pressure-tested in the flow direction (with a positive surface pressure applied to the well), whereas, in an inflow test, a negative pressure is created across the barrier by lowering the pressure above a barrier to verify if the barrier can hold back the formation pressure. This chapter covers three types of PPTs used for verifying the cement integrity:

✓ The conventional PPT is conducted by applying pressure to verify integrity of all barriers, including cement, following cementing operation and before shoe drill-out.

Well Cement Integrity
DOI: https://doi.org/10.1016/B978-0-443-51564-4.00007-2

✓ Leak-off test (LOT) is conducted after shoe and some formation drill-out to verify cement integrity around the casing shoe and determine the maximum allowable mud weight for the next whole section.
✓ Formation integrity test (FIT) is carried out with a similar purpose as LOT.

Sometimes, it is not possible to properly test a barrier by applying enough pressure from the surface. Thus conducting a PPT across a barrier may not be practically possible. In such a case, an NPT or inflow test should be alternatively conducted as a replacement test method. Sometimes, it may be necessary to validate results of PPTs, just for a verification purpose. In this case, the inflow test would be conducted as a complementary test method. Inflow tests are generally regarded as more reliable than pressure tests, especially for liner laps. There have been many field cases where PPTs failed to detect leaks, but inflow tests successfully identified them. In Macondo Deepwater Horizon well the crew could not properly use inflow tests' results to identify the kick and blowout.

An inflow test involves reducing the hydrostatic pressure in the well to create a net differential pressure from the formation into the wellbore (API, 2013). This test is used when positive pressure cannot be applied upstream of the barrier being tested. It is essentially the reverse of a PPT, where an underbalance is created within the casing/well across the barrier. The underbalance can be called a pressure drop, or "negative pressure differential," or "vacuum." The well condition is then monitored to detect any potential flow of formation fluids. In practice the underbalance is created by lowering the hydrostatic pressure above the barrier, achieved by replacing the fluid above the barrier with a lighter one. It is recommended to conduct this test enough time after cementing is completed, allowing the cement to stabilize in its in situ conditions. For a liner string, this period typically aligns with the time required for drill-out operations to clean the cement above the liner lap or previous casing.

Inflow tests are advised to verify the integrity of the following barriers in the well:

- Cement sheath behind the shoe-track of a casing string or liner string (including the shoe and float collar) and the wellhead assembly, including casing hangers (for subsea wells). It is important that test be performed before drilling out the shoe track.
- Plugs (either mechanical or cement plugs) installed for plug and abandonment (P&A).
- Plugs set in the well before initiating workover operations or removing the Christmas tree.

Therefore, in the last section of the chapter, following description of different components of a standard inflow test tool, several methods available for performing the test are presented. Next, preparation measures and operational procedures are explained and presented stepwise. Next the analysis of different test results and final interpretation are presented stepwise to infer about well integrity. Finally, two real case studies in the ME and GOM are presented to clarify the materials given.

The material on inflow testing presented in this chapter is the most comprehensive, user friendly, straightforward, and practical text material ever, to the best of author's knowledge. Ashena (2025) presented some material on inflow tests, but the current chapter's material is more complete and more useful for industry and academia, delivering many practical guidelines.

6.2 Conventional positive-pressure test

In this work, a conventional PPT refers to a test conducted after wait-on-cement and before drill-out of shoe-track cement. In this test, hydraulic pressure is applied in the well/casing with the DP/casing annulus being closed by a blowout preventer (BOP). The test pressure should reach the maximum anticipated surface pressure, which may exceed the formation pressure. A PPT should be continued for at least 5 minutes, at low- and high-test pressures, with possible pressure changes over time being monitored and documented. Via this test, it would be possible to verify if surface barriers (wellhead seal assembly) and downhole barriers (casing and shoe-track cement) can contain pressure inside the well without failure.

6.2.1 Procedure:

The procedure to conduct a PPT (with a surface BOP) for cement integrity is

1. Stop displacing cement when bumping is observed.
2. If a company's policy requires getting cement return, slurry return at the surface should be already observed at the flow line prior to observing bumping. In such a case, when you stop the displacement, it is essential to first open the side valves of the casing spool and subsequently wash/clean with water the path from the flow line to the side valves.
3. Set the casing hanger in the wellhead spool to seal the annulus around the casing string.
4. Close the side valves.
5. Connect the safety valve on top of the casing string. Then, nipple-up the Kelly (in the case of using a Kelly system). Therefore the setup will be ready to pump mud through the drill string for well safety, let's say to defend the well against possible hazards.
6. Unset the slips, which are hanging the casing string in the rotary table.

7. Close a BOP, for example, the annular or the pipe rams.
8. Apply pressure up to, for example, 2000 psi in the hole via a kill line (i.e., at the rig-floor). Stop pumping and immediately close the kill line valve.
9. Monitor the surface/annular pressure at the choke manifold.
10. Record and check the pressure trend versus time for 5–15 minutes (depending on company's policy). Then, proceed to interpret the records/results.

6.2.2 Interpretation:

In the case of not observing any pressure drop, the PPT is inferred to be successful (no leakage). However, in the case of observing any pressure drop:

a. First check the surface wellhead (spool and casing slips/hanger) in the case of any leakage.
b. If no wellhead leakage is detected after checking the surface:
   i. For liners the existence of a liner lap's leak may be inferred.
   ii. For casings the leakage could be most likely at the shoe-track (particularly the shoe) because of bad cementing or a hole in the casing string.

6.2.3 Actions:

If no pressure drop is observed during the test, indicating a successful positive test (PPT described above), it does not necessarily mean there is no leakage through the liner lap; thus it is still required to conduct the inflow test (to be discussed in the next chapter). However, in the case of a pressure drop observed during a PPT, it means there is leakage or hydraulic communication, including through the liner lap or somewhere else in the well. The following actions should be taken:

1. *Squeeze cementing:* Remedial squeeze cementing may be decided as the first action to resolve hydraulic communication through the liner lap. To be able to squeeze cement slurry:
   a. The drill pipe (DP) is run with the retrievable test treat squeeze (RTTS) packer assembly.
   b. Set the packer.
   c. Pump the wellbore/drilling fluid and check if enough fluid injectivity into the leakage can be taken. Getting enough injectivity of at least 3 bbl/min is an important prerequisite for cement injection. *Note*: In the case of an unsuccessful PPT, there is still probability of getting enough injectivity.

d. Following analysis of the injectivity test, cancel the operation if the injectivity is too low. Alternatively, if injectivity is great enough, proceed to design squeezing parameters, including the slurry volume and the depth of the RTTS packer.
   i. The slurry volume should be less than the volume of rat hole between the liner lap and bottom of the DP or squeezing assembly to prevent cement from getting stuck.
   j. The packer depth should not be too near the liner lap depth to prevent possible cement sticking.

2. *Tie-back:* In the case of not getting the required injectivity, it signals that the leakage is very small and squeezing is not possible. As the second measure, tie-back of the liner is usually proceeded.

6.3 Leak-off test

As the name indicates, the LOT determines the pressure at which the formation leaks off, or let's say the fluid leaks into the formation. Leak-off occurs just before the formation gets fractured by the hydraulic pressure. Not only does the determined pressure give an idea of formation fracture pressure and the maximum mud pressure that can be applied to the wellbore during drilling the next-hole section, but also the integrity of cement at the shoe or the shoe-track is verified. In other words a successful LOT confirms the strength of the cement bond around the casing shoe. Note that the casing shoe is usually considered the weakest point in the wellbore ("loss" zones could be encountered as the well is drilled deeper), and its cement integrity is of greatest importance.

As a prerequisite for getting meaningful results from LOT, it is important to have accurate mud weight. To have these conditions, it is recommended to condition the well using bottoms-up and use representative mud samples for measuring their weights. In addition, accurate pressure data records require using calibrated pressure gauges.

As for execution of the tests, following cementing, drill-out the cement shoe-track and then drill 10–20 ft of the formation rock, shut in the well/annulus using the BOPs. Using pump truck, pump into the drill string very slowly (e.g., half a barrel per minute) and record the stabilized casing or annular pressure and the pumped volume. Continue this until an increase in casing pressure is observed. At this point, stop pumping for a while, until the casing pressure stabilizes, and then record this pressure. Continue pumping at the next increased volume and follow the same procedure until all the pressure and volume points are recorded and plotted. When a sudden pressure drop is observed, it indicates cement failure or a formation fracture at the shoe. When the beginning of a deflection from linearity is observed in the plot, it means the leak-off has occurred, with

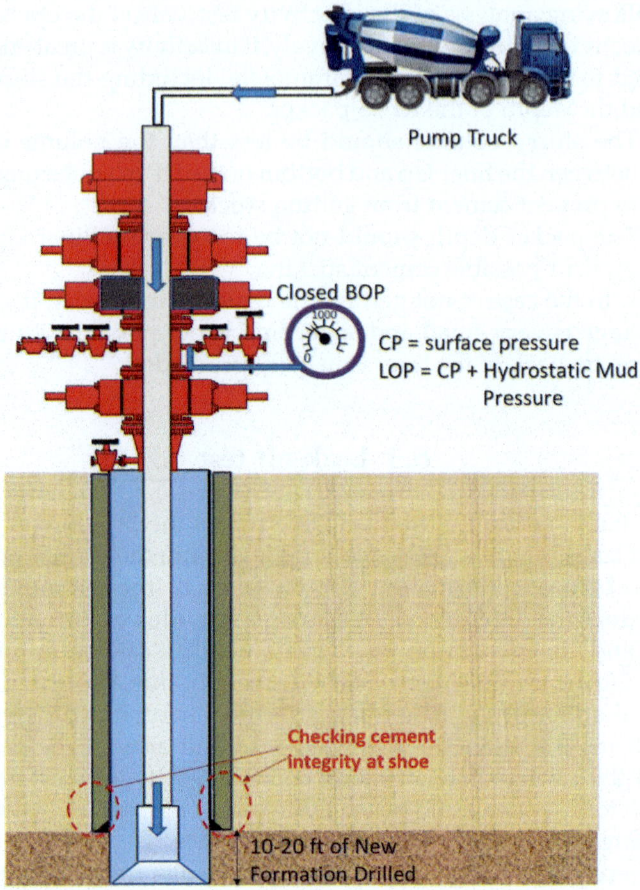

FIGURE 6.1 Schematic of the set-up for performing a leak-off test. *BOP*, Blowout preventer; *LOP*, leak-off pressure.

the surface/casing pressure being called leak-off pressure (LOP). Fig. 6.1 shows the test schematic with the BOPs closing the annulus. Fig. 6.2 shows an example plot of a case study LOT with the surface LOP of 750 psi. The in situ LOP indicates the maximum strength of shoe's formation and cement integrity, which is found by adding the surface LOP to the hydrostatic mud pressure:

6.4 Formation integrity test

An FIT is usually preferred over an LOT because of (1) possible pressure restrictions of downhole equipment (such as casing strings or surface equipment such as wellhead) or (2) low formation fracture pressure. The

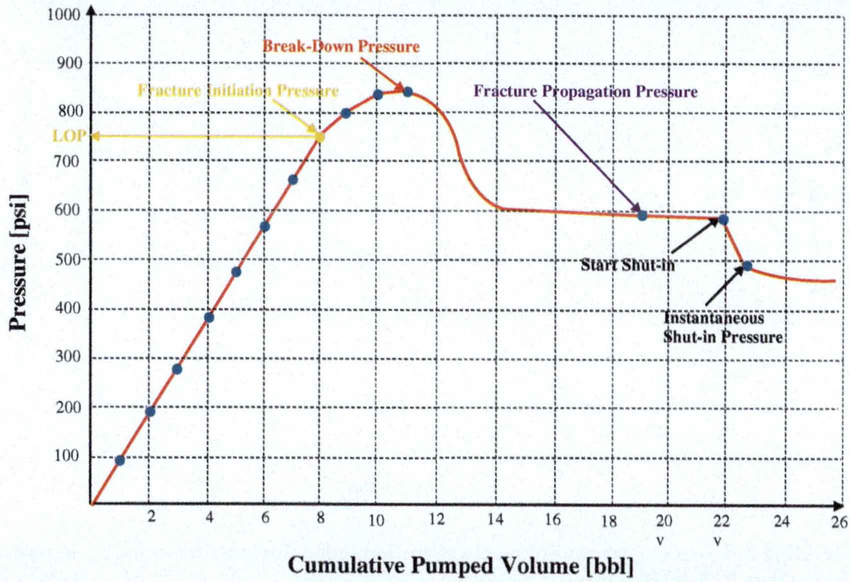

FIGURE 6.2 The results of a field example leak-off test for the hole section below the 13 3/8-in casing at the depth of 3300 ft. The surface leak-off pressure is identified as 750 psi. This is the point after which immediate deflection occurs.

first reason category should not usually be the case, as during the design it is expected that equipment with high enough pressure rating be selected. Anyway, in this case, the same procedure as the LOT is followed, but with the difference that pressure is applied up to a predetermined magnitude. At this point the integrity of cement and formation integrity is verified. There is no intention to break the formation, but rather FIT is to ensure that whole-section target depth can be safely drilled without integrity issues. In the case of sudden pressure drop during FIT, lack of cement or formation integrity at the shoe is inferred. Fig. 6.3 shows the output pressure curve of a field example FIT for the 8.5-in hole (below 9 5/8-in casing) at the depth of 9300 ft (2834 m).

6.5 Inflow tests

An inflow test is performed after cementing and before casing shoe drill-out. It creates an underbalance across the barrier by lowering the hydrostatic pressure above it, typically by replacing the fluid with a lighter one, and monitoring for any potential flow of formation fluids. However, several previous publications, such as Dahi and Santos (2023), have offered academic insights into inflow testing. API RP 65-2 (2010) just briefly

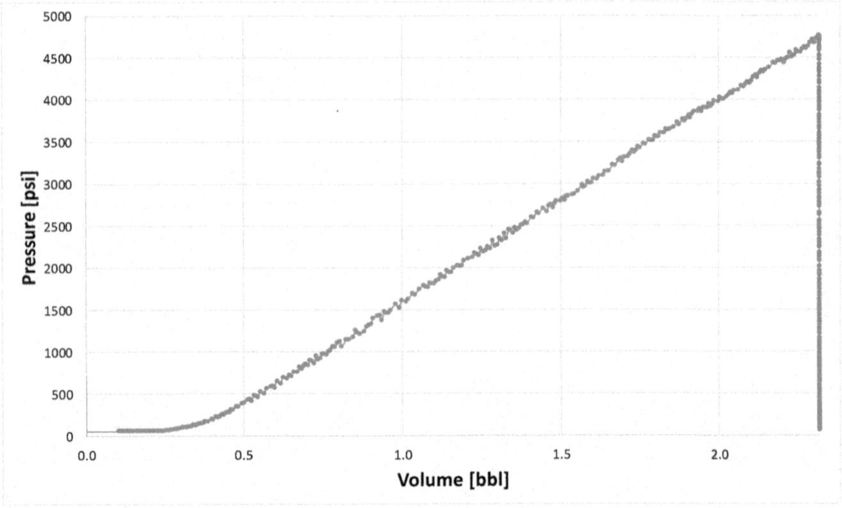

FIGURE 6.3 A field example of results of the formation integrity test in the 8.5-in hole at the depth of 9300 ft (2834 m).

mentions inflow testing or negative pressure testing without giving further details.

This section provides both a theoretical and applied/practical explanation and comprehensive discussion of inflow tests, including the associated testing tool components, methods, procedures, and the interpretation of pressure profiles.

6.5.1 Testing tool components

An inflow test is carried out using the similar tool and a similar method as is used in the drill stem test (DST). An inflow test is sometimes called "dry DST" by some practitioners because this test involves no or minimal production of hydrocarbons (oil or gas) to the surface, and it is used mainly for pressure and integrity evaluation. As there is minimal or no fluid production, the test duration is usually shorter than regular DST. The inflow test is recommended for verifying cement integrity behind the liner, or in other words, to detect if there is or is not any leakage in the liner, particularly at the liner lap. Inflow testing is the best way for this detection because CBL tools cannot contribute because of their limitations in double- or multiple-pipe cases (e.g., where a liner string overlaps a casing string).

To conduct this test the assembly should be carefully prepared. The sequence of the components (from top to bottom) is shown in Figs. 6.4 and 6.5. They are listed as follows:

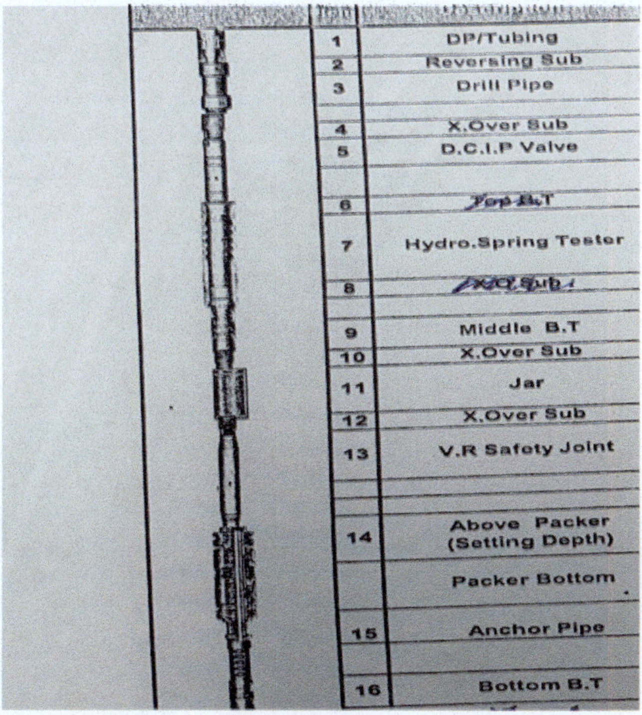

FIGURE 6.4 The components of the inflow tool in a case study.

1. *DPs:* DP joints constitute the means of extension of the inflow test bottomhole assembly (BHA) to the surface.
2. *Reversing sub*: This sub is activated, at the end of the flow and the closing periods, to conduct reverse circulation for displacing the cushion water in the string with the mud in the annulus. To enable that the ports of this sub should get essentially opened by dropping a bar from the surface (see Fig. 6.6).
3. *DP joint*: A single DP joint is placed between the reversing sub and the dual close-in valve [downhole continuous injection pressure (DCIP)], which is useful for sampling purposes. The sample would allow characterizing the properties of the formation fluids.
4. *Crossover sub*: It is a sub that connects the DP joint to the DCIP valve.
5. *DCIP valve*: The dual close-in valve (DCIP) is a downhole valve that is initially open when it is run. It is kept open during the open/flow period (usually 3 hours). However, it must be closed, by 11 times string/rotary table rotations, for static bottomhole pressure measurement during closed/shut-in (also called buildup period). The position of this valve is switched by rotating the string/rotary table a few times.

FIGURE 6.5 Schematic of the inflow test assembly.

Note: Sometimes, in inflow tests, it may be asked for two open and two closed periods, as is the case with regular DST. DCIP valve can alternatively be opened and closed just by making 14 times string/table rotations; it is possible to provide the required number of open and closed positions. In the design of the DCIP valve, the maximum allowable number of rotations possible by the valve is 55. For predicting possible failure at the reverse circulating sub, as a contingency measure, there are 16 rotations predicted to enable the last opening of the valve and allow reverse circulation through DCIP.

6. *Hydro-spring tester (valve)*: When the tool is run in the hole until it reaches the bottom, this downhole valve is run in closed position because its spring is in stretched condition, keeping it closed. After setting the packer the valve should be opened to begin the flow period.

FIGURE 6.6 (A) Reversing sub and its ports and (B) the bar used to shear the pins and open the ports.

FIGURE 6.7 Hydro-spring tester valve (which is above the packer).

To do that, weight is put on the assembly to compress the spring to make its bottom ports open. Fig. 6.7 shows a photo.

7. *Middle BT*: The middle test sub (MBT) is sub where a digital sensor (which is more reliable than the mechanical bottom BT) is placed inside (with some distance from the bottom above the packer). The sensor measures the hydraulic pressure every few seconds (say 20 seconds). This sensor measures the pressure profile in the middle of the well, typically located between the surface and the bottom of the well. Note that the maximum working hours of the pressure gauges are either 48 or 72 hours. Fig. 6.8 shows a photo.
8. *X-over sub*: It is a crossover sub connecting the middle BT to the jar.
9. *Jar*: It is used to release the string in the case of getting stuck.
10. *X-over sub*: It is a crossover sub connecting the jar and VR safety joint.

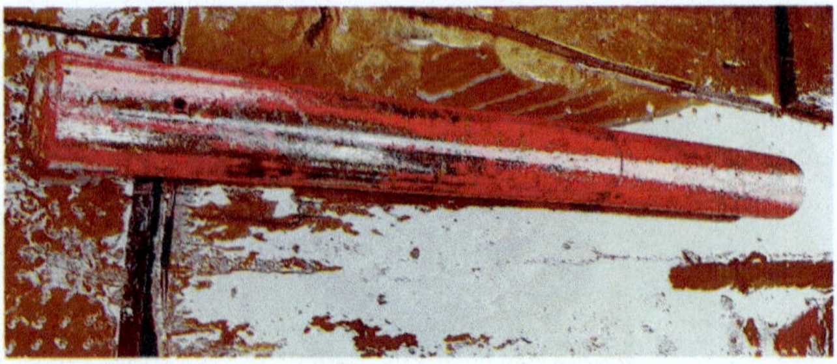

FIGURE 6.8 Middle BT (MBT) sub or container (where the sensor is latched inside).

11. *VR safety joint*: In the case of assembly getting stuck, it is used for mechanical back-off of the upper inflow test tool. To activate the VR joint, first, the crew would rotate clockwise until the maximum allowable tool's rotations for changing the DCIP valve's position are over. This means that from that time on, turning the string clockwise would not affect the DCIP valve position but rather would open the VR safety joint.
12. *Packer*: The packer, when set, would allow removal of the hydrostatic pressure column from the bottomhole. Therefore, if there are any leakages, the wellbore fluids can come into and be detected in the inflow test assembly. Fig. 6.9 shows a photo.
13. *Anchor pipe*: It is a perforated pipe to let pressure equalize during the run-in-hole operation.
14. *Bottom BT*: The bottom test sub (BBT) is sub where a sensor is placed to measure the bottomhole pressures every 20 seconds. Fig. 6.10 shows a photo. This sensor measures the pressure at the bottom of the well (bottom hole) during the test. The bottom refers to the deepest part of the drill stem, typically where the test is conducted to evaluate the reservoir conditions. Note that the maximum working hours of this pressure gauge are, like the MBT, either 48 or 72 hours. The two MBT and BBT sensors are used for calibration of the results.

Note: The pressure gauges must be already calibrated prior to being brought to the rig site.
Note: The pressure gauges/BT in the inflow test assembly are the same as the regular DST. Just, in regular DST, there is an additional pressure gauge between the DCIP valve and the hydro-spring tester.

6.5.2 Methods and procedures

This section presents methods, preparatory measures, and operational procedures for conducting inflow well tests.

FIGURE 6.9 The packer used inflow test (picture taken during retrieval).

FIGURE 6.10 The bottom BT/BBT sub (where the bottom BBT pressure gauge is inserted inside).

There are two methods for inflow testing:

6.5.2.1 Method 1—Assembly and packer

As for the first method, Fig. 6.11 illustrates a typical well schematic of inflow test for land drilling across a liner string's shoe-track (and particularly the liner lap). Similarly, for subsea drilling, it is possible to inflow test using a packer to be run with the inflow test assembly, which is like the land drilling method, just the offshore BOPs must be closed unlike the

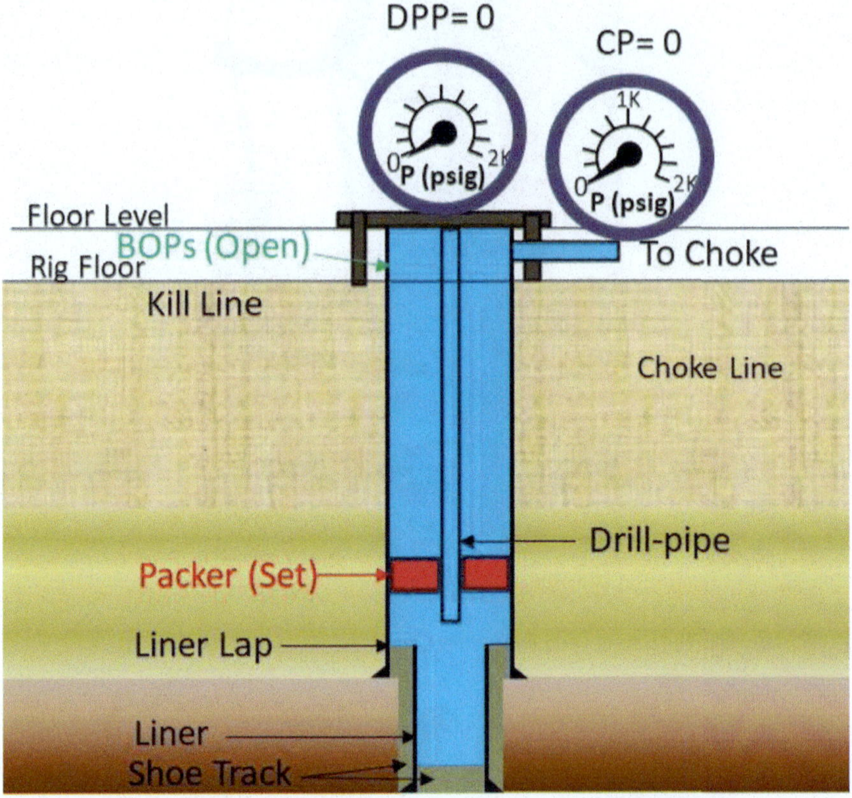

FIGURE 6.11 "Method-1 with assembly and packer, land": The well schematic of a successful inflow test with a land drilling rig across the liner string, including its shoe-track and lap, with the shoe-track not yet drilled-out. *BOP*, Blowout preventer.

land test (see Fig. 6.12). The preparation and procedure for conducting this method is presented in the following part of the section. Note that the first case study in this chapter follows this method.

6.5.2.1.1 Preparation

There are some preparations required prior to commencing the inflow test as follows:

✓ A mill or drilling bit assembly is run to drill-out cement in the previous casing. This cleaning is needed because the packer of the inflow test tool should be set in the lower section of the previous casing string. In some cases, it may be necessary to run another mill or bit in order to clean the remaining cement debris above the float collar of the casing or the latch-down collar (LDC) of the liner.

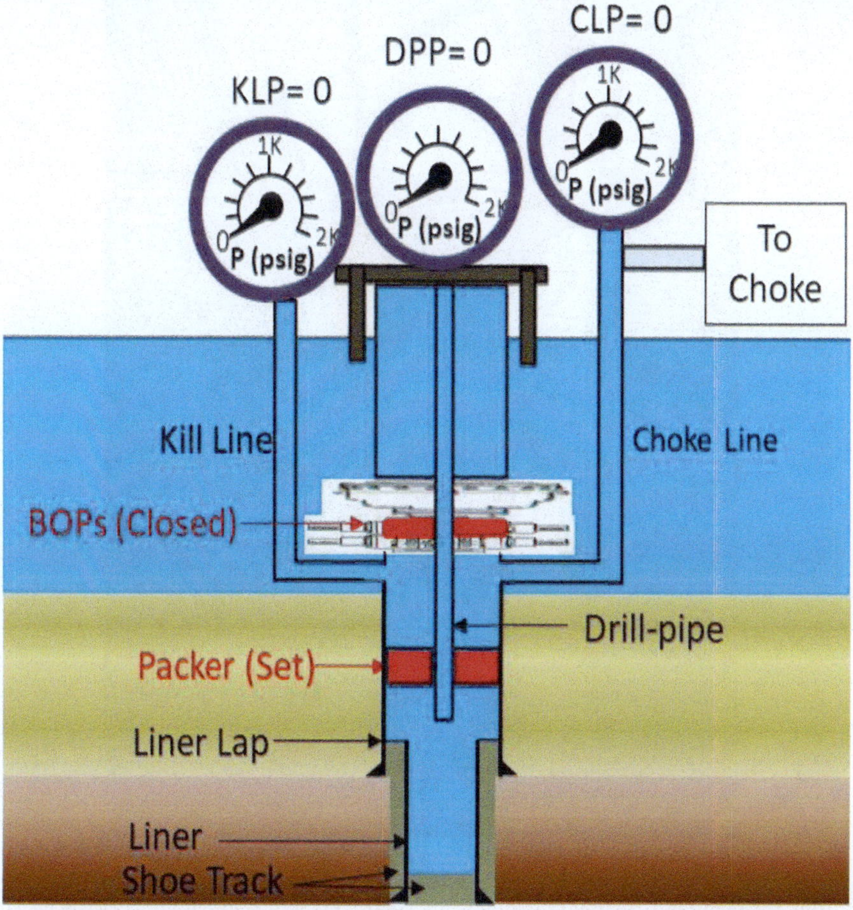

FIGURE 6.12 "Method-1 (with assembly and packer, a subsea well)," well schematic of a successful inflow test with a subsea drilling rig across the production liner string, including its shoe-track and lap, with the shoe-track not yet drilled-out. *BOP*, Blowout preventer; *KLP*, kill-line pressure.

- ✓ The components of the inflow test BHA must be already transported and prepared at the rig-site.
- ✓ A choke manifold is necessary at the rig floor to be connected to the DP/string for monitoring possible flow out from the DP in a controlled way. Fig. 6.13 shows a photo. Connect the small choke manifold at the floor to the main rig's choke manifold for further safety.
- ✓ From the manifold a needle valve is connected to a slim hose that is stretched to a bucket full of water (to observe any possible bubbles during the open/flow period). The manifold itself is connected to the shaker box so that if the cushion water was suddenly and forcefully

FIGURE 6.13 Choke manifold with a needle valve to observe possible bubbles flow during the open/flow period (3 hours) using connectors and Chiksan pipes fasteners. The head is connected to the manifold and from there to the water bucket.

pushed by the formation fluids, the fluids can be diverted to the shaker box and tank. Note that a safety valve is connected to the top of the string at the surface, which we can close at any emergency time.

✓ Enough cushion water should be predicted. Knowing the estimated formation pressure, it is essential to determine the required cushion water's height above the DCIP valve in the assembly to create enough underbalance above the barrier, say 1000–1500 psi.
  - A specified height of cushion water is maintained above the closed valve, with the remaining space in the DPs being empty (air).
  - In the first case study to be presented in this chapter the cushion water height was 1390 m (~4560 ft) to create 1500 psi underbalance pressure.

✓ A risk assessment should essentially be performed prior to inflow testing, to evaluate the potential risks associated with the test.

6.5.2.1.2 Operational procedure

Using the DPs (say 5 in) the BHA is run in the hole near the barrier (say the liner lap). The bottom of the assembly is closed because one of the assembly's valves is run closed (i.e., the hydro-spring tester valve), whereas the other upper valve (the DCIP valve) is closed. Continue running until the distance of the packer from the liner lap is say 20 m (~66 ft). At this point, running is stopped. Next the packer is set by a few times or more clockwise rotations of the string (minimum two times for each 1000 m/3281 ft) as well as putting weight. The rotation and weight would convert the packer from the un-J to the J position such that packer's dice are forced to stick into the casing. The packer should be pressure tested to say 3000 psi depending on requirements and conditions. Setting the packer would reduce the hydrostatic wellbore pressure that is measured downhole. At this point the driller puts some weight on the string so that the hydro-spring valve is activated/opened. Therefore opening of the hydro-spring valve creates a hydraulic conduit enabling communication from the bottomhole to the DP. With the underbalanced condition created the crew wait for the open/flow period of 3 hours, during which the pressure at the bottom of the packer is measured by the bottom sensor (called BBT) while the pressure at the bottom of the packer is measured by the middle sensor (MBT). During the open/flow period the top of the string is already connected to a manifold at the rig floor. This manifold may be connected to the rig's choke manifold for more safety. In case there is a very large liner lap leakage, which may cause pushing the cushion water up suddenly by the formation pore fluids, it is recommended to close the DCIP immediately (before the 3-hour period) to stop the flow. Such an observation would indicate that the liner has a critical lap integrity condition.

After the open/flow period (3 hours) the rotary table (and the testing tool assembly) is rotated for 11 times clockwise so that the DCIP valve gets closed so that it would be possible to investigate the well in closed/shut-in condition.

At the end of the operations the cushion water must be circulated out first. To do this the shear pins of the reversing sub (the component above the DCIP valve) are sheared so that its hole can appear. This state would allow the crew to make a reverse circulation for reverse circulating the cushion water out. At the end of circulation the well condition is expected to be balanced or equalized in pressure. At this point the packer is unset (by applying some overpull force), which would let the hydrostatic pressure increase at the bottom of the tools. Following packer unset the tools are retrieved back to the surface. In the retrieval, first the reversing sub (which is located above the DP joint) is disconnected. However, the other components are separated from the bottom (starting from the bottom BT). During the retrieval the pressure sensors are retrieved; that is, their digital or mechanical memories are detached and taken for interpretation. In addition, a fluid sample is taken from the hydro-spring tester.

Note: The inflow test is made using the DST assembly. In the regular DST, two open periods and two closed periods are required. However, in the inflow test, it is usually required to have only one open/flow period and one closed period, which is considered the operational difference between the inflow test and regular DST. However, company policy for inflow testing may ask for two open and two closed periods.

As a summary the procedure is presented stepwise as follows:

1. Run with DPs the inflow test assembly into the hole to a specified depth above the barrier (e.g., liner lap). The assembly is composed of a closed hydro-spring valve, open DCIP valve, and a packer.
2. Set the packer (by applying weight and rotation) at least at 50 ft (15.2 m) above the interval to be tested to isolate the annulus from the DP. Test the packer from the annulus with say 1000 psi pressure.
3. *The 3-hour open-I period*: Open the (hydro-spring) valve in the inflow test BHA assembly (by applying weight on the assembly). This signals the beginning of the 3-hour open/flow period. Monitor pressures and flow through the DP. It is expected that flow and pressure will drop to zero after a while.
4. *The 3-hour close-I period*: Close the DCIP valve (by rotating the table/string 11 times clockwise) and monitor pressures and possible flow.
5. Repeat steps (3) and (4) to check the well flow conditions and well barrier integrity twice (*3-hour open II* and *3-hour close II*). To switch the position of the DCIP valve each time, rotate the table/string 14 times clockwise.
   Note: The BOPs are open during the test; therefore it is possible to monitor the level of mud in the annulus. Any drop of the mud level signifies a lack of integrity in the packer or drill string.
6. Unset the packer and reverse circulation (1 hour).
   a. To prepare for reverse circulation of cushion water at the end of the test, first, a bar must be dropped into the well to break the pins in the reversing sub above the DST tool. Once the shear pins are broken, this can be confirmed by a drop in the mud level within the annulus. Next, immediately seal the annulus/well by closing the annular preventer.
   b. Do reverse circulation of the cushion water by pumping mud from the annulus through the kill line, with returns flowing from the drill string back to the choke manifold via the installed Chiksan pipes.
   c. Analyze the return fluids at the choke manifold.
7. Unset the packer and pull the string out of the hole.

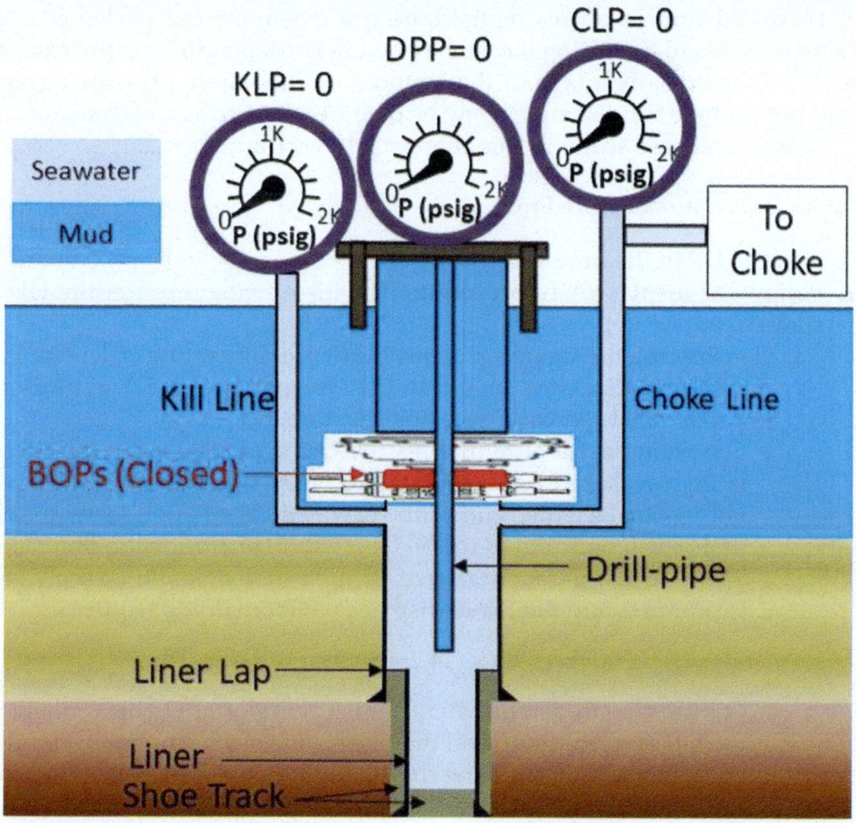

FIGURE 6.14 Method-2 (displacement), well schematic of a successful inflow test with a subsea drilling rig across the production liner string, including its lap, with the shoe-track not yet drilled-out. *BOP*, Blowout preventer; *KLP*, kill-line pressure.

8. The pressure sensors are detached and taken for pressure analysis. Take a sample of downhole fluid encapsulated from the hydro-spring tester (valve) and in the string.

6.5.2.2 Method 2—Fluid displacement

The second method is less standard than the first one, but it is sufficiently acceptable to be applied to either land or subsea drilling. In this method the wellbore fluid should be essentially lightened or displaced by a lighter fluid. For subsea drilling, fluids inside the kill line(s) and choke-line(s) must be displaced as well. To create enough underbalance ranging from 1000 to 2500 psi depending on the company's policy, the right length of DP is run in the hole, and then the heavy drilling mud in the hole is displaced by (light) seawater. Next, observe the surface flow and pressures. Fig. 6.14 shows a well schematic and test mechanism.

As an advantage of this method, no test assembly and packer would be required, and operational cost may be lower (depending on the case as well). As the disadvantage of the method, no downhole pressure curves and bottomhole fluid samples may be obtained for further analysis.

The second case study in this chapter follows this method.

6.5.2.2.1 Operational procedure

1. Run the DP in the hole and
2. Follow the displacement procedure. The displacement procedure takes some steps.
   a. Considering the rig pump's maximum pressure rating, calculate the depth steps that crew should stop at during running DP in the hole to pump seawater for displacing the mud.
      i. For example, for a pump pressure rating of 5000 psi, if the circulating pressure loss is assumed to be 2000 psi, there is 3000 psi left to displace the mud with seawater. Calculation leads to the required length of DP (h) of 10,395 ft (3168 m) for the first step: Thus 116 stands (each 90 ft) of DP are run first, and then seawater is pumped into the DP to displace wellbore fluid.

$$3000 = 0.052 \times (14.2 - 8.65) \times h \rightarrow h = 10,395 \text{ ft}$$

      ii. For the second displacement step, it is important to also consider the DP displacement, and therefore, in a fewer number of stands, crew should stop running the DP to pump seawater. The calculation is not provided here for conciseness.
      iii. Continue the procedure above until the same volume of mud is returned from the riser.
   b. Use a spacer fluid between the seawater and heavy mud to allow separation (not allowing them to mix).
   c. Continue this until all the mud is far above the BOPs in the riser.
3. (Only for subsea) Pump seawater into the choke-line(s), boost-line, and kill line(s) to displace their fluids into the marine riser. This action will replace the fluid in the said lines (mud or whatever fluid) with seawater.
4. Then, close the BOPs. This action will isolate the riser and its hydrostatic pressure from the wellbore.
5. Monitor for at least 30 minutes (possible) flow at the surface (top of DP) or kill line as well as pressure at the surface. For good integrity, it is expected that there will be no flow and no pressure observed at the surface.

6.5.3 Analysis and interpretation of test results

To interpret an inflow test in terms of well barrier integrity, there are several sources of information and data that must be analyzed in combination.

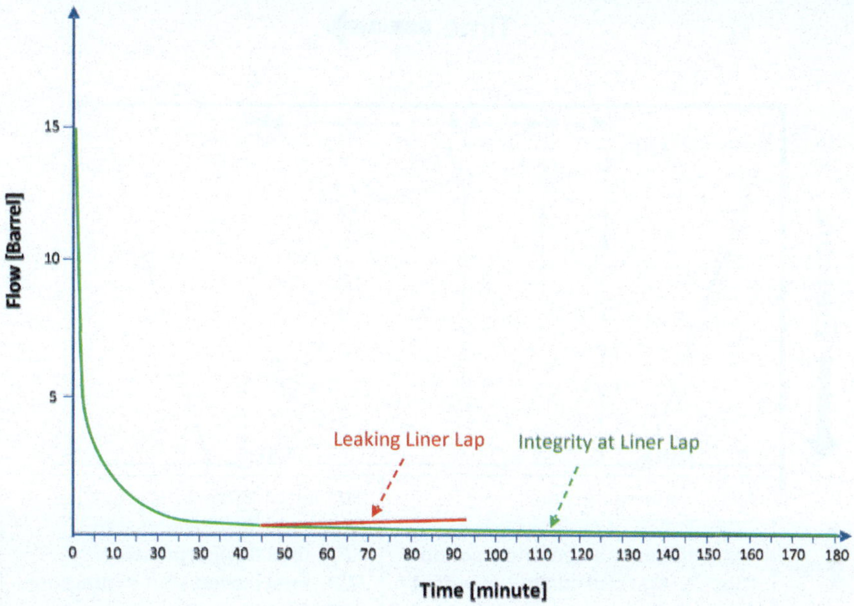

FIGURE 6.15 Example results for successful and unsuccessful inflow tests during the open/flow period.

An inflow test is considered successful (negative) when the following six observations are used to confirm integrity of well cement and liner:

1. *Flow observation at the surface*: If the flowout is zero or minimal but is decreasing very rapidly toward zero, it would indicate a good well integrity on an inflow test. In case strong flowout/medium or strong bubbles are observed in the water bucket at the surface, we infer lack of well integrity and connection with the formation fluids. The minimal flowout is because of thermal expansion effect because in deep-hole tests, results are adversely affected by thermal expansion and contraction of the long fluid column in the wellbore or by rig movements in subsea rigs. To counteract the thermal expansion effect, it is recommended to allocate enough test observation time to ensure a reliable test. It should last at least 3 hours for each open/flow or close periods or even longer as per the client's requirements (see Fig. 6.15).
2. *Pressure readings at the surface*: When there is good integrity, it is expected that zero surface pressures are measured at the DP side as well as the annulus side. However, because of thermal expansion, some minimal pressure may show and read at the surface, but it should decline to zero after a while.
3. *Downhole pressure curves*: Following an inflow test, the pressure and temperature curves are downloaded (from the retrieved pressure gauges)

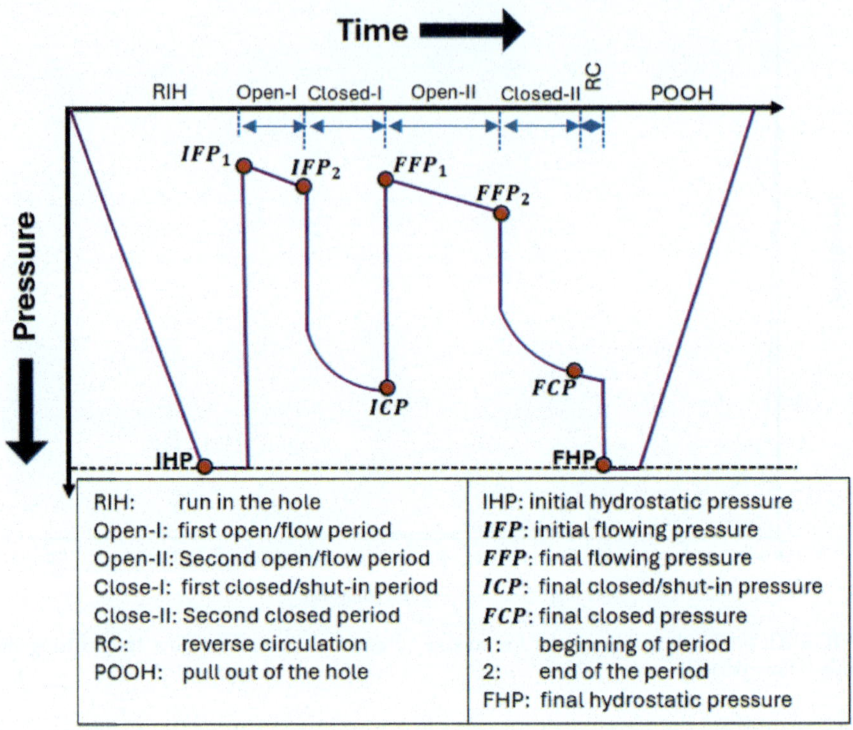

FIGURE 6.16 Stages and points in downhole pressure readings of a typical regular drill stem test.

and analyzed. An inflow test is considered successful (good integrity across the barrier) when it is observed there is no or minimal pressure increase ΔP at the end of the closed-in period (that is the difference between the final flowing pressure and initial closed pressure). Fig. 6.16 displays different stages and pressure points in a regular DST with two open and two closed periods, whereas Fig. 6.17 presents pressure curves during a typical inflow test.

a. Some industry practitioners consider the minimal pressure rise as "$\Delta P < 120$ psi."

b. Some others consider "$\Delta P < 150-200$ psi" for deep wells where thermal expansion effect plays more role on pressure rise. Lack of integrity is inferred if "considerable pressure rise" of greater than 200 psi is observed.

c. The difference between ΔP_{mid} and ΔP_{btm} should not be considerable, though it is acceptable to be within 30 psi.

d. The final hydrostatic pressure (FHP) is expected to be equal to the initial hydrostatic pressure (IHP) to confirm sensors' calibration and good operation.

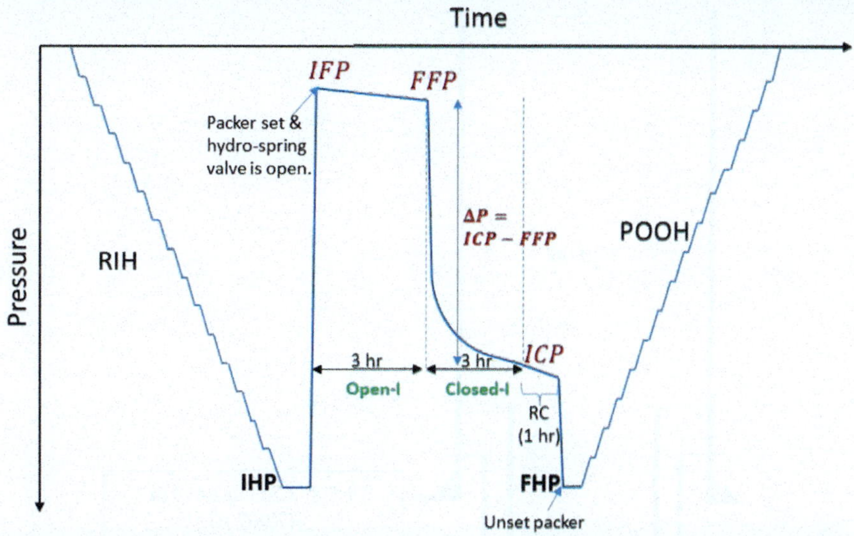

FIGURE 6.17 Stages and points in downhole pressure readings of a typical inflow test. Note that final flowing pressure (FFP) in inflow test is the same as IFP_2 (initial flowing pressure at the end of flow period) in a regular DST. *FHP*, Final hydrostatic pressure; *ICP*, initial closed pressure; *IFP*, initial flowing pressure; *IHP*, initial hydrostatic pressure.

4. *Pumped volume versus empty volume*: After reverse circulation, record how many pump strokes (or the equivalent mud volume) are required for the cushion water to be received at the surface. Compare the pumped mud volume with the empty volume (in the drill string above the cushion water). In case the pumped volume is less than the empty volume, it means that there is a formation fluid flow/gain because of possible liner lap leakage. If it is the opposite, it probably indicates some wrongdoings or miscalculation in the execution of the test.
5. *Fluid taken during reverse circulation*: A way of detecting possible hydraulic communication is to measure weights of fluids displaced to the surface during reverse circulation. After receiving the cushion water at the surface, fluid weights are measured, and if there is a light weight observed or traces of hydrocarbons noticed, it can infer lack of integrity. The fluid analysis is a rough indication of well integrity, but it is a quick one because it is useful during reverse circulation before waiting for pulling the string out of the hole and retrieval of the cell fluid sample bottomhole.
6. *The cell fluid sample*: this fluid is more representative of the bottomhole fluids. If there is good well integrity, the fluid encapsulated in the cell between the hydro-spring and DCIP valves should not have hydrocarbons. If there is hydrocarbon found in the sample, it signals hydraulic

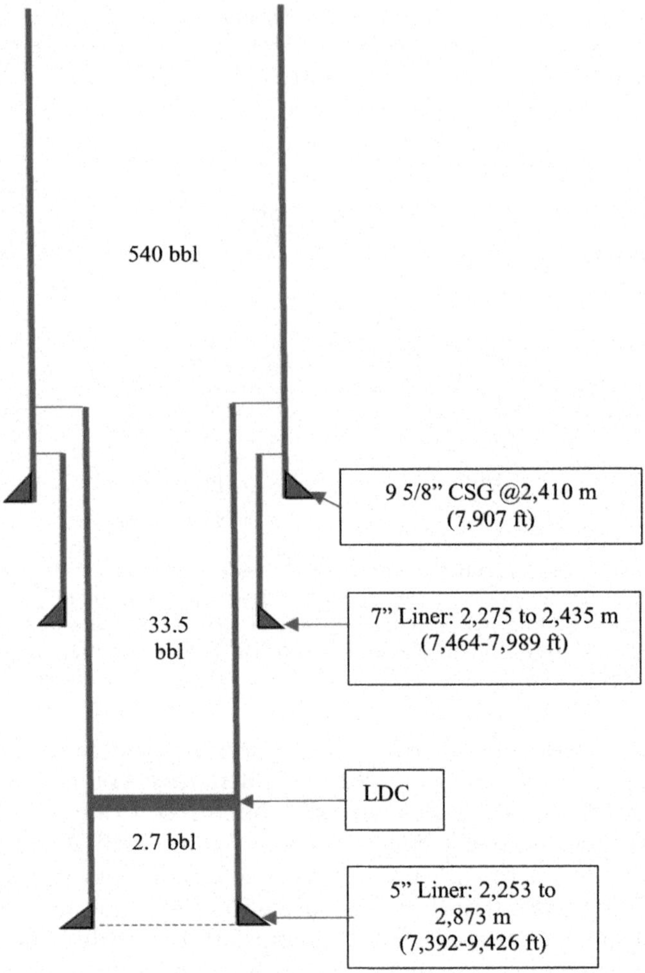

FIGURE 6.18 Wellbore schematic for inflow testing above the 5-in liner lap. *LDC*, Latch-down collar.

communication from the formation during the test, probably through the liner lap.

6.6 Case studies and interpretations

6.6.1 Case study I (Middle East, 5-in liner at 9426 ft)

Fig. 6.18 shows the schematic of a case study well in the ME. After running the 5-in liner string (from 2253 m/7392 ft to 2873 m/9426 ft), it was planned to inflow test the liner, including its lap. Following the tool

TABLE 6.1 Main pressure reading of the inflow test for the case study well.

First (and only) stage	*Note:* mud weight = 9.224 ppg	Middle BT (MBT) pressure gauge @ 2230.8 m/7319.25 ft	Bottom BT (BBT) pressure gauge @ 2235.8 m/7335.66 ft
	IHP (psi)	3507	3515
	IFP_1 (psi)	2016	2054
	IFP_2 (final flowing pressure) (psi)	2016	2054
	ICP (psi)	2070	2097
	FHP (psi)	3507	3515
	Underbalance pressure (IHP–ICP) (psi)	1437	1418
Sensors' calibration	Criterion-1: IHP = Calculated HP	**Yes**	**Yes**
	Criterion-2: IHP = FHP	**Yes**	**Yes**
	Interpretation	**Calibrated**	**Calibrated**
Integrity	Criterion-1: $\Delta P = \text{ICP} - \text{FFP} < 120$ psi	**(56 < 120 psi)**	**(43 < 120 psi)**
	Criterion-2: $\Delta P_{mid} - \Delta P_{btm} < 30$ psi	**Acceptable**	**Acceptable**
	Interpretation	**Good**	**Good**

FHP, *final hydrostatic pressure;* ICP, *initial closed pressure;* IFP, *initial flowing pressure;* IHP, *initial hydrostatic pressure.*

retrieval the downhole pressure memory gauges (bottom and middle BTs) were downloaded and analyzed.

Prior to any interpretation of the results, sensors' performance must be checked as shown in Table 6.1. Both sensors' calibration and proper working are confirmed because the FHP and the IHP are equal to the calculated hydrostatic pressure. For example, for the bottom sensor (BBT), with the mud weight of 9.224 ppg, the hydrostatic pressure is 3517 psi, "$\text{HP} = 0.052 \times 7335.66 \times 9.224 = 3517$," exactly the same as the sensor reading.

Next, for verifying the integrity of liner and its lap, the following observations were analyzed as criteria:

Observation 1: Flow observation at the surface: If the flowout is zero or minimal but is decreasing very rapidly toward zero, it would indicate a good well integrity on an inflow test. In case strong flowout/medium or

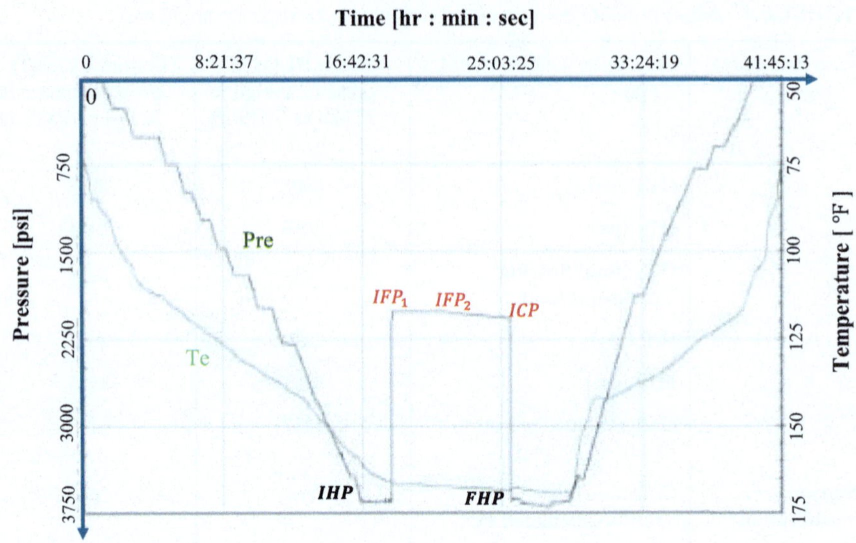

FIGURE 6.19 The downhole pressure and temperature readings of the middle memory gauge (MBT) for testing the 5-in liner, including its lap, showing good integrity. The gauge was digital. Note that the pressure axis increases downward. *FHP*, Final hydrostatic pressure; *ICP*, initial closed pressure; *IFP*, initial flowing pressure; *IHP*, initial hydrostatic pressure.

strong bubbles are observed in the water bucket at the surface, we infer lack of well integrity and connection with the formation fluids.

Observation 2: Pressure readings at the surface: Zero surface pressures were measured at the DP side as well as the annulus side.

Observation 3: Downhole pressure curves: Figs. 6.19 and 6.20 show, respectively, the middle and bottom pressure curves during the entire test, with the important pressure points magnified. Fig. 6.19 shows the temperature curves as well, showing that temperature rises to the maximum at the end of the closed-shut-in period. Table 6.1 shows the important pressure points for analysis and interpretation of integrity of the well liner and cement I.

The pressure differences (ΔP) between the ends of the flowing state and the closed/shut-in states were evaluated by subtracting FFPs from the ICPs. They were 56 and 43 psi (for the middle and bottom sensors), which were ideally less than 120 psi, indicating proper integrity. The ΔP for the middle and bottom sensors should not be so much different, though a slight difference of 16 psi (<30 psi) is acceptable. In this case study, there is integrity in the liner and its lap (no leakages).

Observation 4: *Pumped volume versus empty volume*: the pumped mud volume during reverse circulation was equal to the empty volume in the DP; it means that there was good well cement and liner integrity.

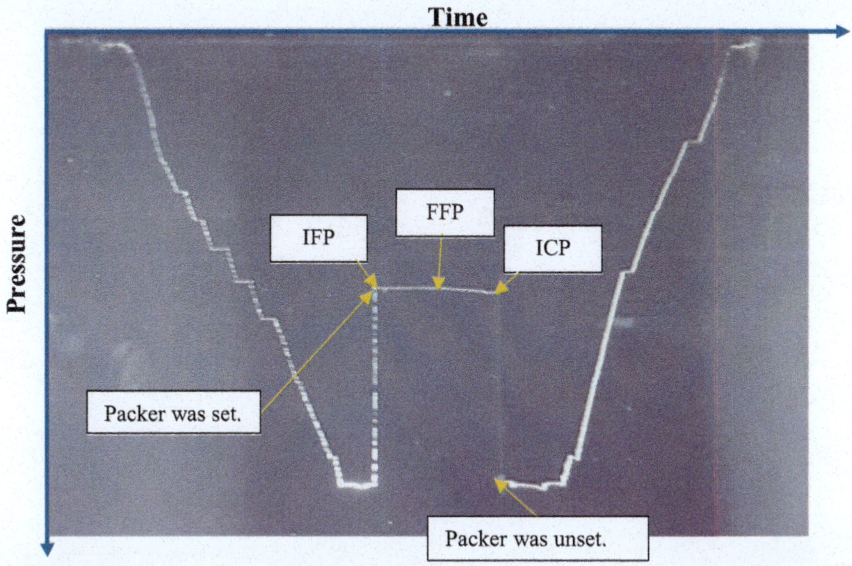

FIGURE 6.20 The downhole pressure readings of the bottom memory gauge (BBT) for testing the 5-in liner string, including its shoe-track and lap, showing good integrity. The gauge was mechanical. *ICP*, Initial closed pressure; *IFP*, initial flowing pressure; *FFP*, final flowing pressure.

Observation 5 and 6: Fluid taken during reverse circulation and cell fluid sample: The fluid sample taken from the hydro-spring tester was weighed by the chemistry test man, and its weight was the same as mud weight (i.e., no external fluid intrusion from the formation).

6.6.2 Case study II (Gulf of Mexico, 7 × 9 7/8-in casing at 18,304 ft)

6.6.2.1 Background

The second case study pertains to the inflow tests performed in BP's Macondo Deepwater Horizon well in the US GOM, which experienced a blowout. Fig. 6.21 displays the holes and casings of the well. The seawater depth was 5067 ft (~1544.3 m). The well was drilled using the 8 ½-in pilot bit and 9 7/8-in reamer to the total depth (TD) of 18,360 ft (5595.85 m). Finally the tapered production casing of 9 7/8 × 7-in casing was run, but only the 7-in portion of the string across the pay-zone. The casing shoe was set at 18,304 ft (~5578.8 m), which was off-bottom with a rat hole of 54 ft (~16.5 m). The formation temperature and pressure were, respectively, 262°F (~ 128°C) and 13,270 psi (or pore pressure gradient of 13.9 ppg or 0.723 psi/ft).

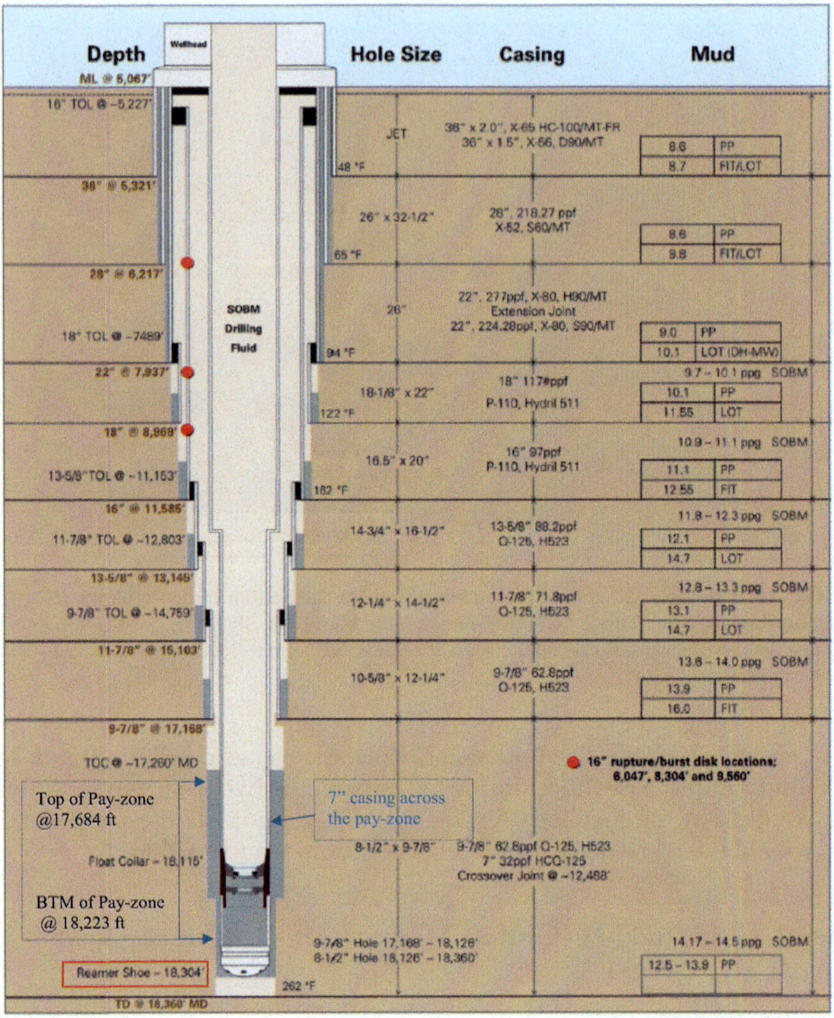

FIGURE 6.21 Well schematic of BP's Macondo well, including the casing strings, hole sizes, and mud weights (*Redrawn from BP. (2010). Deepwater horizon accident investigation report. BP*). The bottom of the pay-zone is at 18,223 ft (5554 m). *FIT*, Formation integrity test; *LOT*, leak-off test.

Note that the mud used in the last hole was synthetic oil-based with the last weight of 14.2 ppg, which means that drilling was completed with 257 psi overbalance pressure. In the last 100 ft of drilling, lost circulation occurred. Lost circulation materials (contained in 400 bbl LCM pill) were spotted in the well. In addition, the mud weight was reduced to 14.2 ppg to solve the lost circulation problem. Full circulation was regained, and the well was successfully drilled to the TD of 18,360 ft (5596 m).

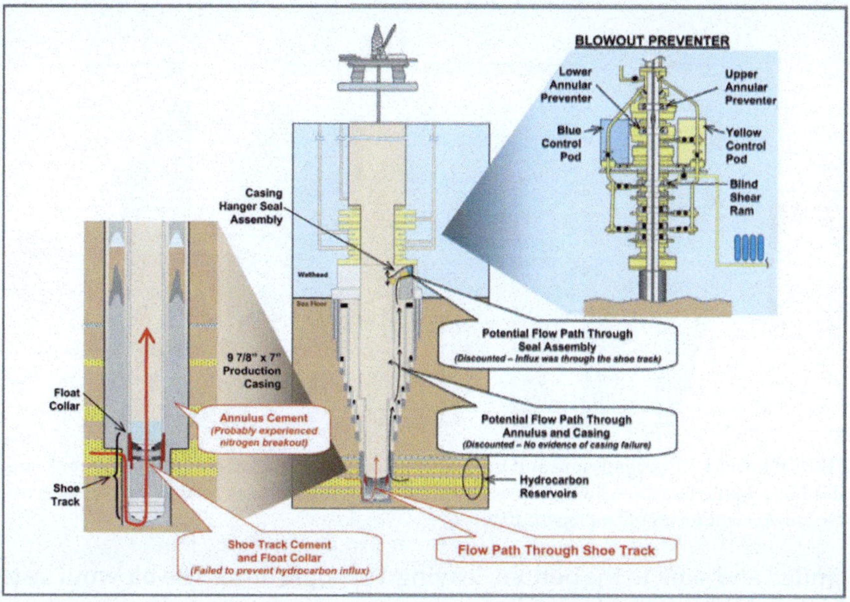

FIGURE 6.22 Well schematic of BP's Macondo well, including gas migration to the surface. *(Redrawn from BP. (2010). Deepwater horizon accident investigation report. BP).*

The production casing was run and cemented. The cement behind the production casing was a nitrified foam class H cement of 16.7 ppg weight, which must not have had enough stability, and thus nitrogen breakout had most likely occurred. There was a long rat hole with mud lighter than cement there, which also caused contamination of the cement slurry. Therefore the shoe-track cement barrier had most likely been contaminated with some channels for hydrocarbon/gas migration (see Fig. 6.22).

About 10.5 hours following completion of cementing the PPT commenced to check casing and shoe-track integrity.

Following successful completion of the PPT, inflow tests (or negative pressure test) were conducted to cross-check the PPT result and verify the ability of the mechanical barriers (shoe-track, casing pipe, casing hanger seal assembly) to withstand underbalance pressure and keep their integrity. However, the crew failed to analyze them correctly and did not recognize the lack of well integrity and loss of well control. With the inflow test having been accepted, the well was returned to an overbalanced condition, preventing further influx into the wellbore. Later, as part of normal operations to temporarily abandon the well, heavy mud was again replaced with seawater, underbalancing the well. Over time, hydrocarbons had already flowed up the production casing and passed the BOP. Following several more mistakes in sequence, the well blew out, and the rig

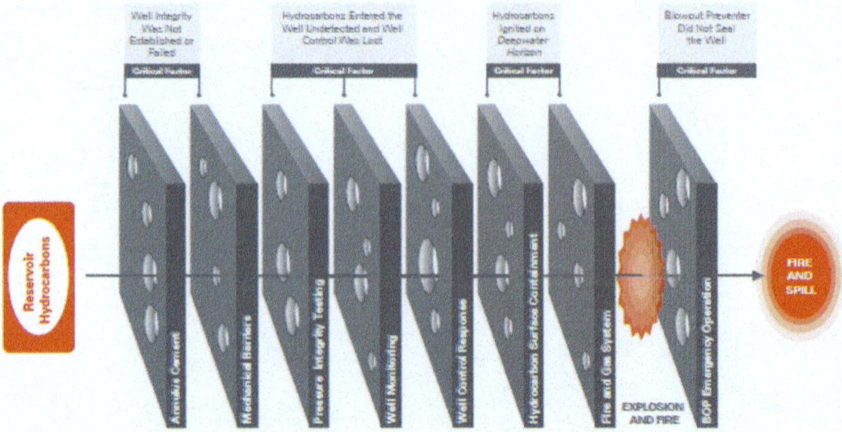

FIGURE 6.23 Poor pressure and integrity testing was configured as one of the root causes and holes in the Reason's Swiss Cheese incident analysis. *(Redrawn from BP. (2010). Deepwater horizon accident investigation report. BP).*

ignited and sank to the bottom, leaving 11 people dead. The blowout lasted 86 days before being controlled.

In the analysis of the blowout that occurred on April 20, 2010, it was revealed that poor pressure and integrity testing was one of the root causes, shown as a slice of the Reason's Swiss cheese (see Fig. 6.23). Therefore these tests are explained and analyzed in this part.

6.6.3 Inflow tests

Two inflow tests were carried out to verify integrity of the cement, shoe-track, and pipe of the 7-in portion of the production casing. The tests were conducted using the second method of inflow testing explained before.

6.6.3.1 Inflow test-1

6.6.3.1.1 Procedure:

In the first inflow test, 8367 ft of DP was run in the hole, and then it proceeded to displace the heavy drilling mud in the hole by seawater, which could create the underbalance pressure of 2400 psi, "$0.052 \times 8367 \times (14.17 - 8.65) = 2400$ psi." The heavy mud was displaced by a 16-ppg spacer fluid (LCM pill) and seawater. The spacer was used to properly separate the seawater and heavy mud.

Next, seawater was pumped into the boost, choke, and kill lines to displace mud. But 1200 psi was left trapped in the kill line, not bled off (BP, 2010).

Thus the LCM pill had been displaced by seawater from inside the DP to be located in the lower part riser (with the height of about 450 ft/137.15 m length) just above the BOPs. But the well was losing fluid downhole

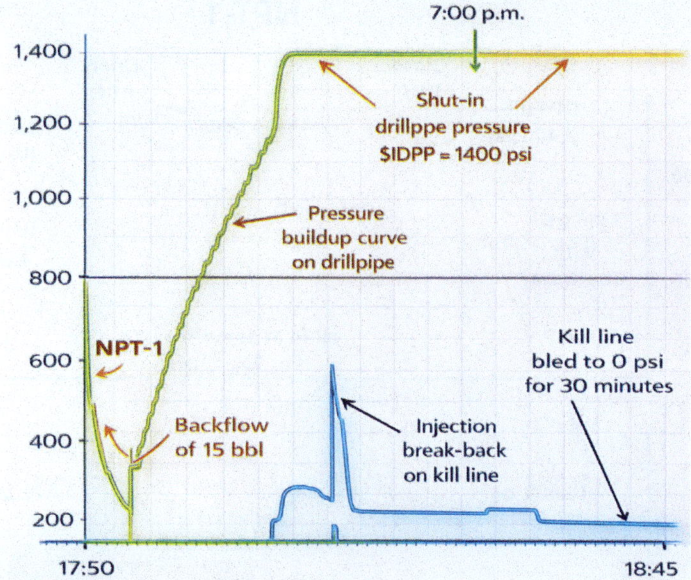

FIGURE 6.24 Surface pressure observations during the first and second inflow tests (NPT-1 and 2) for BP's Macondo well. *(Redrawn from Turley, J. A. (2014). An engineering look at the cause of the 2010 Macondo blowout. In Paper presented at the IADC/SPE drilling conference and exhibition March 2014. https://doi.org/10.2118/167970-MS).*

during, about 65 bbl, so this amount of 16 ppg LCM pill was below the BOP, some above the end of the DP at 8367 ft, and the rest below the DP (reference US CSB Macondo Investigation Report Appendix 2a). The annular preventer was closed to isolate the hydrostatic head of the fluids in the riser from the well. However, the preventer did not seal around the DP, resulting in more of the heavy spacer leaking down past the preventer. The LCM materials played a minor role in the first inflow test, but they had a significantly negative effect on the second inflow test.

Observation/Evidence 1: The crew displaced most of the mud with seawater and dropped the bottomhole pressure for about 2200 psi, but when the pressure struggled to drop the final 200 psi for 5 minutes (until 18:00 hours as shown in Fig. 6.24), sudden flow of 15 bbl of seawater was observed from the DP (BP, 2010; CSB, 2016). With about 2200 psi underbalance created by the inflow test and because of contaminated cement inside and outside the 180-ft shoe-track, the pay-zone was enticed to flow down the casing shoe, as well as the shoe-track and then across the float collar. The float collar did not seal or had malfunctions most likely because of load damage, or alternatively, its two flappers had not been well converted to one-way check valves because of insufficient flow rate. In any case the surface flow is the first concrete evidence that there was absolutely no integrity.

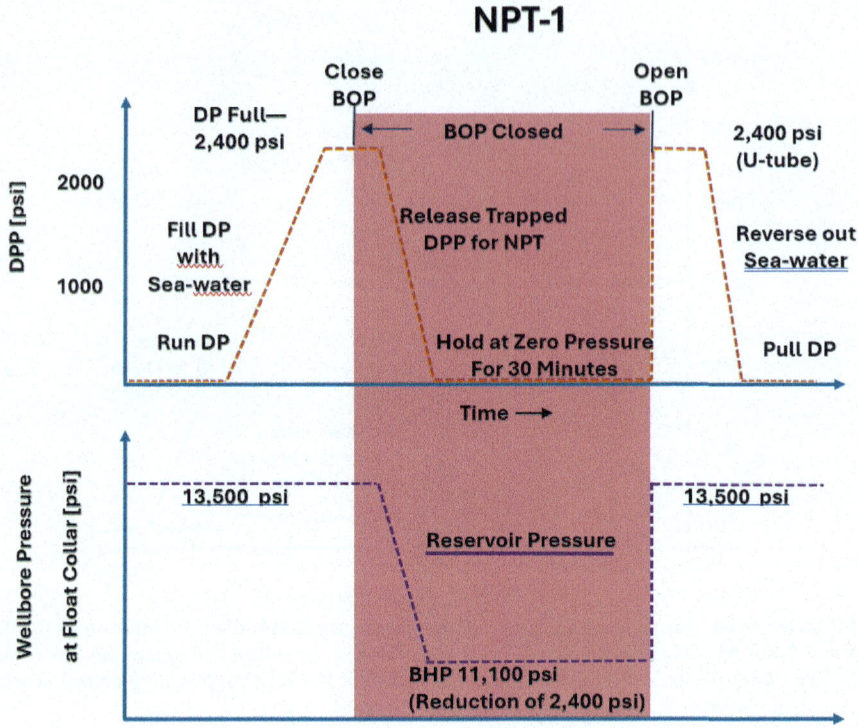

FIGURE 6.25 Expected surface DP pressure curve for the first inflow test (NPT-1) in the BP's Macondo's well (*Redrawn from Turley, J. A. (2014). An engineering look at the cause of the 2010 Macondo blowout. In Paper presented at the IADC/SPE drilling conference and exhibition March 2014.* https://doi.org/10.2118/167970-MS). *BOP*, Blowout preventer.

Observation/Evidence 2: Following the flow at 18:00 hours the DP pressure started building up and reached 1400 psi after ~35 minutes, shut-in DP pressure SIDPP of 1400 psi. The rate of pressure increase was ~40 psi/min. This is the second clear evidence of lack of integrity or hydraulic communication with the reservoir formation. The crew ignored all these pieces of evidence for lack of integrity, and instead of taking action to control the well, aborted the first inflow test, assuming a mistake had happened, and decided to run another inflow test for verification. In addition, 1200 psi left trapped in the kill line, which could not be bled off, is another evidence.

Observation 3: Downhole pressure curves. Such curves were never found because the well blew out (and no downhole pressure recorders were run). But, even if they were used, they could provide no useful information for integrity interpretation because there is no closed or shut-in period because the test was aborted at the beginning of the open/flow time. Fig. 6.25 shows the expected surface DP pressure curve as well as the downhole pressure curve for the first inflow test (NPT-1) in BP's Macondo well. Of course, as the inflow test was conducted using method-2, no downhole pressure could be recorded, and the figure just displays expected hypothetical curves.

6.6.3.2 *Inflow test-2*

The second inflow test was carried out this time using the 5000 ft (~1524 m) kill line instead of the DP. The surface kill-line pressure (KLP) showed 500 psi for 3 minutes (which was ignored) and zero for the mandated 30 minutes with no flow observed at the surface. These observations were false because the LCM materials or possibly gas hydrates (Vavik, 2020) had plugged the kill line bottom, but the crew was not aware of this and did not guess its possibility. Therefore they declared the well was secure. In less than 2 hours after the conclusion of the second inflow test, the well blew out.

As described above, some LCM pill had unwantedly stayed below the annular BOP, so they were prone to entering the kill line when the second inflow test started. The LCMs (or gas hydrates) apparently caused plugging of the kill line during most of the second inflow test; therefore no fluid flow was observed at the top of the kill line, though, misleading the crew.

There was a 3-minute injection break-back in Fig. 6.24 showed there was a clear hydraulic communication path from the 3 1/16-in kill line into the wellbore (and into the pay-zone), still unaffected by the wad of LCM spacer pill in the BOP. With the kill line being hydraulically connected to the reservoir, about 500 psi KLP is read at the kill line. This pressure is a very clear indication/observation of lack of integrity or kick/hydraulic communication, which was also ignored by the crew.

There was still a remaining 100 psi KLP at the end of the second inflow test that was bled for 30 minutes. The bleeding of KLP occurred because the narrow kill line was entirely blocked by the LCM material (or gas hydrates), whereas the kick fluid continued rising in the well.

No real flow was observed at the kill line. But, during the short duration of the test, no conventional water bucket method was used at the top of the kill line to detect any possible bubble flow.

At 7:55 p.m., the second inflow test was approved, and well was considered secure.

At 9:49 p.m. (less than 2 hours later), the well blew, out signaling the beginning of a long catastrophe. All these signify a series of mistakes made just for the inflow tests in this well.

References

API. (2013). *API RP 96: Deep water well design and construction* (1st ed.). API.

American Petroleum Institute (API). (2010). Isolating Potential Flow Zones During Well Construction (API Recommended Practice 65-2, 1st ed.). Washington, DC: American Petroleum Institute.

Ashena, R. (2025). *The practical handbook of well control*. Routledge, Taylor and Francis Group ISBN 9781032753850 https://www.routledge.com/The-Practical-Handbook-of-Well-Control/Ashena/p/book/9781032753850?srsltid=AfmBOooZqOJyBoNQ8l WhWyMcNMmdX4-St1TUa2goxJzlN_6-31cO4lBo .

Ashena, R., Bahreini, H., & Lotfi, M. (2023). Statistical analysis of past kicks and blowouts occurred in a Middle Eastern oilfield. *Journal of Petroleum Exploration and Production Technology, 13*, 2063–2082. https://doi.org/10.1007/s13202-023-01664-9.

BP. (2010). *Deepwater horizon accident investigation report*. BP.

Dahi Taleghani, A., & Santos, L. (2023). *Well integrity from theory to practice*. Springer ISBN 978-3-031-19024-7 https://doi.org/10.1007/978-3-031-19024-7 .

Holand, R. (2017). Loss of well control occurrence and size estimators, Phase I and II. Report no. ES201471/2. Office/Division Program. TAP. Project Number. 765. Category. Deepwater.

Turley, J. A. (2014). An engineering look at the cause of the 2010 Macondo blowout. In *Paper presented at the IADC/SPE drilling conference and exhibition* March 2014 https://doi.org/10.2118/167970-MS .

CSB. (2016). *Macondo blowout and explosion report*. U.S. Chemical Safety Board https://www.csb.gov/macondo-blowout-and-explosion/.

Vavik, D. (2020). *Implications of gas hydrates in drilling and completion—A scientific root cause analysis of the deepwater horizon disaster*. NTNU.

CHAPTER

7

Temperature noise and fiber optic logs

Jacqueline Greenwell, Rahman Ashena

Smart Energy Data Analytics (SEDA)-Group, Houston, TX, United States

7.1 Introduction

"Despite its potential, the cement bond log is probably one of the most abused, misused, and misunderstood logs used in the oil field today. Miscalibration, inadequate information, and a severe lack of standardization are enough to push petroleum engineers into a morass of bewilderment." ***(Fertl, Pilkington, and Scott "A Look at Cement Bond Logs" JPT, June 1974)***

The above quote from 1974 is still applicable to petroleum engineers today. Interpretation issues with acoustic bond log tools may still occur, but by combining this data with temperature logs, noise logs, and the continual advancements in technology, a more comprehensive understanding of the cement barriers' quality and, overall, well integrity status can be achieved.

Cement placed in the annulus is the most common barrier element utilized in the prevention of flow behind casings, between formations, or to surface. If cement is designed and placed successfully according to acceptance criteria and well conditions, and if cement bond quality from logs run in the wellbore is evaluated as providing sufficient coverage and height to hydraulically isolate zones as per requirements, why does the industry continue to observe sustained casing pressures (SCPs) and encounter hydrocarbons and aquifers with the ability to flow to surface or cross-flow behind pipe?

A cement bond log is typically run postcementing during drilling operations and prior to any further operations such as fracture stimulation, acidizing, perforating, and completion activities. In some instances a cement bond log may not have been performed on a well, or historical records are lost. Postcementing activities during drilling have the potential

Well Cement Integrity
DOI: https://doi.org/10.1016/B978-0-443-51564-4.00003-5

to compromise cement integrity. Similarly, production-related activities, such as injection (CO_2 or water flood) or well cycling and thermal loads, may also impact the cement sheath's structural ability to hydraulically isolate zones.

Once a well is completed and put into production, running a standard sonic or ultrasonic cement bond tool may not be feasible because of limitations such as cost, tubing inside casing, reduced internal diameter, obstructions, and presence of multiphase reservoir fluids or gas. Although there continue to be developments and advancements in tool technology, until that time, alternate technologies and diagnostic methods will be required to ascertain the location and flow path of cement integrity breaches.

The SCP, measured from a side outlet annulus valve at surface, which is not attributed to thermal effects, may manifest once the well begins producing. Refer to Chapter 4 "SCP—Prevention, Monitoring, and Remediation," for more information. When there is SCP in an annulus, and when bled down, the pressure builds up to a similar value; this alone may indicate a cement integrity problem. This indicator is readily observable at the surface. However, the absence of SCP does not necessarily indicate that there are no cement integrity issues in other vertical sections below the top of the cement (TOC). The cross-flow between one underground formation or zone to another is not necessarily detectable with surface pressure monitoring. Cross-flow behind casing can occur and impact productivity, cause formation damage and scaling, contaminate aquifers, or pressure up a depleted zone because of an undetected leakage pathway.

During the abandonment phase, it may be necessary to confirm that no behind-pipe integrity issues exist before permanently abandoning the well, to comply with standards or regulatory requirements. The capital investment required to reenter a failed abandonment can be substantially greater than the initial abandonment cost, present technical challenges, and potentially increase risk exposure. Enhanced knowledge and comprehension of the well and cement integrity by the engineering team before planning an abandonment job can result in lower overall costs and risks, thereby improving the likelihood of a successful permanent abandonment.

Although a cement bond log aims to provide information on the quality of the bond to casing and formation, as well as TOC and cement interfaces, there may be limitations to evaluating the complete cement integrity profile and robustness with any degree of confidence (Ashena et al., 2014). The following examples of cement bond log imperfections may be reasons for a company to investigate additional technologies for cement barrier verification:

- Old or legacy wells with limited data or no bond logs,
- Highly deviated wells may potentially have no bond log,
- Limits with vertical resolution, depth of investigation, as well as radial coverage,
- Quality control while logging providing uncertain data, for example, eccentricity, tool out of calibration, incorrect parameters set by logging company,
- Types of cement being evaluated, that is, foamed or lightweight cements,
- Variations in borehole fluids; gas ingress,
- Cement not set, or log was run too soon after cementing,
- Microannulus presence with no (surface) pressure applied during logging,
- Incorrect or incomplete interpretations; lack of required data supplied,
- Thin cement sheaths, and
- Poor log presentations of data to clients.

Note that there may be other subsurface integrity issues or leakages that are related to cement or complement cement failure. For example, there may be tubing, packer, and/or casing leaks. These leakages could also be picked up by noise and temperature logs. A comprehensive set of data, combining information from cement bond data, noise and temperature logs, as well as lithology and reservoir data, will facilitate a holistic analysis that is more accurate and straightforward. Therefore this chapter will cover some supplementary tools and technologies to complement the cement bond log data, or lack thereof, or to provide a more holistic view of the well cement integrity from not only the drilling phase but also to enable continued evaluation and verification of cement as a barrier that may have altered during completion, intervention, and production activities, and any significant zonal isolation problems prior to abandonment phase. Therefore this chapter covers the following topics:

- Traditional logging tools
  - Temperature logging
  - RTD (resistance temperature detector) and thermistors
- Noise logging
  - SNL (spectral noise log)
- Distributed fiber optic (FO) sensing (DFOS) categorized as:
  - DAS—distributed acoustic sensing, and
  - DTS—distributed temperature sensing.
- Other Indicators
  - Tracer logs

TABLE 7.1 Performance response time of resistance temperature detector (RTD) in flowing water versus air (recreated from Lee and Lee, 2019).

Temperature response time (second)			
Flowing water (0.4 m/s)		**Flowing air (1 m/s)**	
50% Response	0.14	50% Response	7
90% Response	0.35	90% Response	21

7.2 Conventional temperature logging

Temperature logs fall into two categories: conventional and FO. Conventional temperature logging tools are utilized to survey a wellbore for vertical thermal changes in the wellbore fluid and have been used for downhole production surveys as early as the 1930s (McKinley & Thornhill, 1994). These logs can provide information such as TOC, detect movement of fluid or gas behind casings, wellbore temperature gradients, and production or injection performance data. The TOC measurement and flow behind casing(s) within an annular space can assist in determining characteristics of the cement position behind the casing.

It is preferable to take a log of temperature data during a down pass (as it is lowered in the wellbore), with the temperature sensor as close to the bottom of the tool string. This is to reduce any agitation of the wellbore fluids (while running the tool in) that may impact the accuracy of readings. For cased-hole logging, it is recommended, as a minimum, to run the temperature survey with a CCL (casing collar locator) for depth correlation and/or a GR (gamma ray tool) for improved depth matching/accuracy if an existing open-hole gamma ray is available.

The sensor requires time to reach equilibrium with the surrounding fluids during logging. The response time will be affected by the wellbore fluid medium and logging speed. Heat transference will occur more effectively in water (fluid) than air (gas) (see Table 7.1).

In addition to logging speed, other factors affecting log accuracy include the sensor technology and calibration state (Sharma et al., 2021). Logging service companies should provide the recommended logging speed for the tool string being run to meet the log objectives required by the client. Tools may be run with real-time surface readouts or as a memory tool where data is downloaded once the tool is at surface.

7.2.1 Temperature sensors

Temperature sensors measure the temperature of the fluid surrounding them; they do not measure the external formation directly. Therefore it is recommended to run temperature logs in fluid-filled wells for the heat transference to occur, as it is much slower to respond in air (Sharma et al., 2021). There are several sensor types that may be used by logging companies:

a. *RTDs*

The sensor of an RTD is made of resistive metal (the conductor), typically platinum, as it is a stable element at high temperatures and has a predictable linear resistance that changes in temperature. Electrical resistance is the ability of the conductor to oppose the flow of electrical current. As the sensor is heated up, the atoms in the conductor vibrate faster and have more energy. This increases the resistance within the conductor as the number of collisions between the atoms and electrons increases, thus reducing the electron flow capability, and the reverse is true with cooling. This resistance is then measured as a change in voltage and provides the output as temperature.

RTDs can be slow to respond to rapid temperature changes but are reliable over a wide temperature range and have better data repeatability.

b. *Thermistor*

Thermistors for logging are also typically made of platinum for stability. Depending on metallurgy, some thermistors can have a nonlinear relationship between temperature and resistance. They have better sensitivity than RTDs; therefore they can respond to faster temperature changes; however, they do not have the same temperature range as RTDs.

Temperature sensor tools are one method of determining TOC and potential leakages across cement barriers; however, they have some limitations. To detect flow behind casing(s) the flow must be sufficient to change the wellbore temperature with a sensor that is sensitive enough to pick up the thermal anomalies; therefore the conventional temperature logs alone may be limited in picking up these minute fluctuations. Al-Sabea et al. (2022) showed a successful case study where temperature noise log was used to determine the source of water in a horizontal well with inflow control devices to segregate nine perforation intervals. Noise logs can be run in conjunction with temperature logs as a complementary method. Noise logging tools have highly sensitive sensors able to identify extremely weak sounds or noises behind the casings and detect if the fluid or gas is flowing through gaps in the cement sheath. In other cases, it may require pumping in fluids and injecting through perforations or down an annulus to detect these changes.

7.2.2 Top of cement determination

When viewing a typical wellbore schematic, often the TOC is provided in the drawing with a perfectly "cement textured" rectangle within a perfectly in-gauge hole. It is important to ask, "From where was this value derived?" An engineer may pick TOC from multiple possibilities; however, the basis of that value should be presented alongside the value in a schematic.

The TOC value may be designated from:

a. Designing cement top from programmed job design (least certainty),
b. Calculating cement top, using integrated hole volume (if available) and volume pumped, taking into account slurry returns and losses,
c. Running a cement bond tool (sonic, ultrasonic, or other), and
d. Run a temperature log during cement setting process.

It is also important that the values are provided in MD (measured depth) and TVD (true vertical depth) if the well is deviated.

Determining the TOC behind casing(s) by temperature logging during the drilling phase can be crucial information for the drilling team to assess any risks because of an unexpected TOC depth:

- The TOC is too low (deeper than expected), which could indicate some cement slurry loss or calculation mistakes. This would require confirming if isolation of mandatory formations has been achieved or not.
- The TOC is too high (shallower than expected), which could indicate kick flow or channeling, or overdisplacement.

This upfront knowledge allows the team to undertake critical rectifications before moving on to the next-hole stage. However, running a log during the drilling phase increases rig time and therefore costs. Risk versus cost should be evaluated, as well as long-term life of well implications, and management of future barrier failures should they occur.

For the abandonment phase the TOC may be critical if there is a requirement to cut and pull casing(s) or perform cement squeeze or circulation behind casing. If the TOC provided was based on design or calculated values, and the actual TOC is higher, the project costs immediately increase, whereas TOC is established either by trial and error or by running a contingent log that was not initially planned.

In addition, the TOC depth could change the operational MAASP (Maximum Allowable Annulus Surface Pressure) values. If the TOC is below the previous casing shoe, the formation strength pressure will enter the calculations. In a closed system where cement top is above the previous casing shoe, formation strength does not need to be accounted for in the calculation. This may significantly change the MAASP value depending on the formation strength data.

In some regions, including the United States, Australia, and many European countries, cementing policies allow cement placement up to a specified depth behind the casing (i.e., not to the surface) provided that hydraulic zonal isolations can be met in accordance with company and industry standards and regulatory obligations. See a typical well schematic

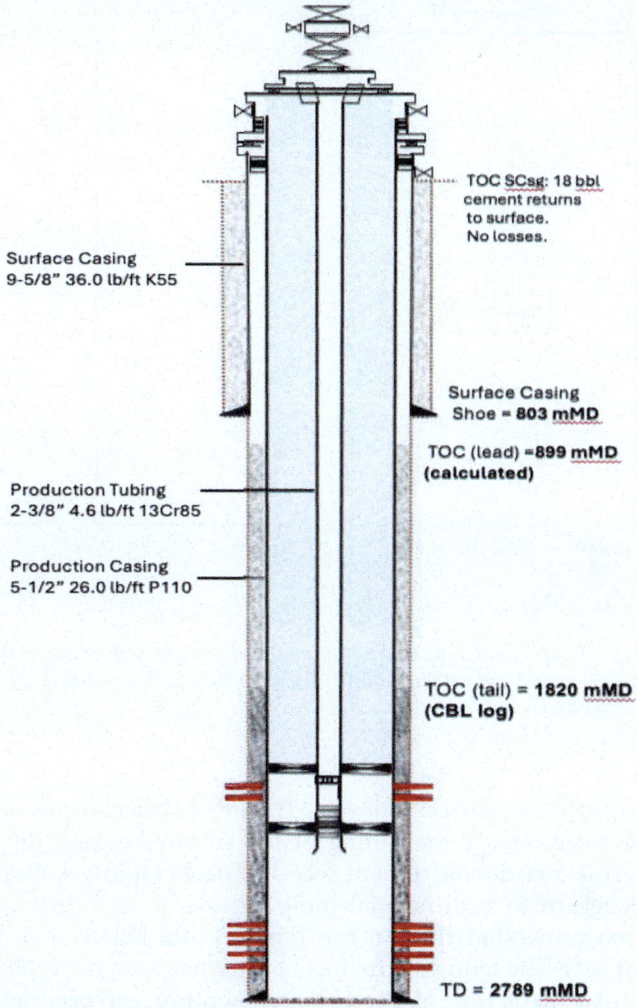

FIGURE 7.1 A simple wellbore schematic showing a typical cement representation. *TOC*, Top of the cement.

shown in Fig. 7.1. Postcementing the operating company can elect to run a temperature log to determine the TOC, but it cannot recognize the cement distribution or vertical isolation between different zones. This operation has been in practice since the 1930s. In other countries and companies, it is required to cement all the casing strings to the surface with the slurry return being observed at the surface. In such cases, running a temperature log was not required.

Following the placement of the primary cement in the annulus, the wellbore fluid temperature increases because of the heat generated by the

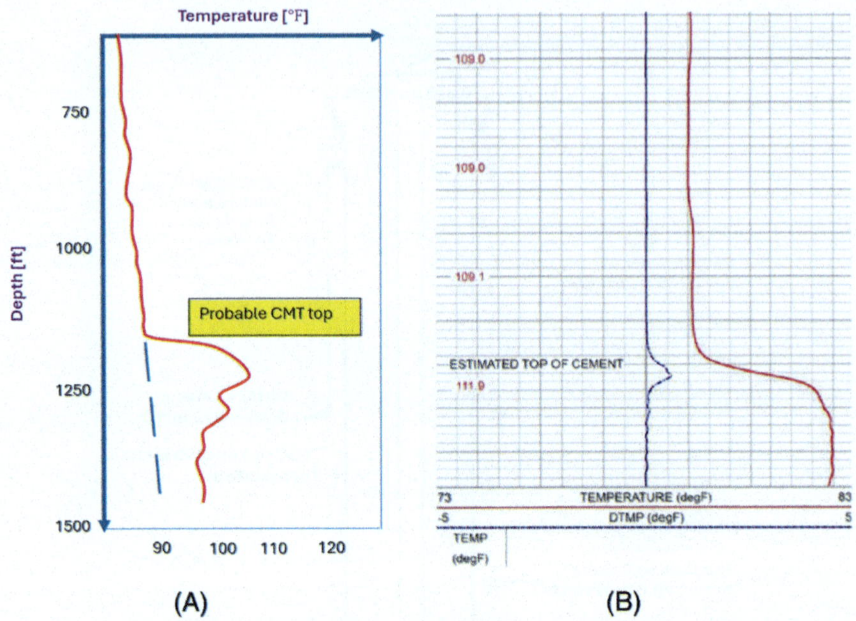

FIGURE 7.2 Temperature logs to determine top of cement: (A) A temperature anomaly is very clear (Allouche et al., 2006), and (B) the detection is clearer using the temperature gradient (Harris, 2021).

hydration process as the cement sets. The most critical factor in obtaining the temperature data is the timing of the survey because the heat from the exothermic reaction as cement sets dissipates rapidly. Laboratory tests indicate a return to near-normal temperatures in 24 hours (Pilkington, 1992). A recommended time to run temperature logs is 4–8 hours after the cement sets. The temperature increase depends on the type of cement, total mass of cement, time elapsed after cementing, and original formation temperature; it can range from a few Fahrenheit up to even greater than 30°F (~17°C) (Harris, 2021). Allouche et al. (2006) mentioned that the temperature anomaly can range from 10°F to 40°F (5.5°C–22°C), with the value depending on the annular thickness and formation thermal conductivity. It is worth noting that lightweight cements tend to have a lengthier thickening time; therefore they generate less overall heat, and the temperature fluctuation may not be clearly detected (Jutten & Morriss, 1990). Fig. 7.2 is a simple temperature log showing the shift in temperature and temperature gradient, providing an estimated TOC during hydration phase. Sometimes the temperature anomaly may not be immediately apparent or visible. In such cases, the temperature gradient can provide valuable insight (see Fig. 7.2B).

TABLE 7.2 Typical fluid flow source detection by noise frequency.

Flow source	Frequency (kHz)
Borehole flow	<1
Completion element flow (perforation)	1–3
Behind casing flow (cement channeling)	3–10
Reservoir flow (matrix flow)	10–30

(Data from Ghalem, S., Draoui, E., Mohamed, A., Keshta, O., Serry, A. M., Al-Felasi, A., Berrim, A., Chaker, A., Hicham, F., Maxim, G., Alfiya, A., Irina, & Aslanyan, A. (2012). Innovative noise and high-precision temperature logging tool for diagnosing complexwell problems. In Paper presented at the Abu Dhabi International Petroleum Conference and Exhibition November 2012. https://doi.org/10.2118/161712-MS).

7.2.3 Conventional noise logging

Noise detection is possible when there is movement of fluid or gas in a turbulent flow status through a sufficient differential. The overall noise generated is from both the internal friction and of the audible high-speed turbulent flow, as well as the audible noise of the rock grain vibrations induced by the flow (Zeinalabideen et al. 2021). Early use of an acoustic recording device was applied by Enright (1955) to detect tubing and casing leaks, and McKinley (1973) continued to build on noise logging as an effective diagnostic tool.

The SNL tool contains a highly sensitive sensor that detects noises within a broad frequency range of around 0.1–60 kHz. It detects both the amplitude and frequency of the noise pattern. Noise is affected by flow rate, differential pressure, and fluid type, whereas frequency depends on the size of the channel opening and reservoir structure (Sadeghi et al, 2024). Table 7.2 shows typical noise frequencies in some flow cases.

The SNL tools can be run in real-time mode or as a memory tool and are commonly run in conjunction with a temperature tool.

The SNL transducer is a highly sensitive piezoelectric hydrophone with an amplifier and data collection module. The piezoelectric quartz crystal converts acoustic pressure waves by compressing (deforming) the crystal into an electrical charge, generating a voltage that is measured.

A limitation of SNL is because of the movement of the tool in the wellbore creating significant background noise, referred to as "road noise," which could mask the acoustic signal of a leak.

To avoid this road noise, SNL should preferentially be run as the station stops; that is, the tool is moved up or down the wellbore to specified stop depths, where the tool remains stationary for a set survey period to establish a reading at that depth before moving to the next station stop. Logging time can be lengthy to obtain the overall objective, increasing cost. Noise logs can then be repeated under different conditions, such as flowing

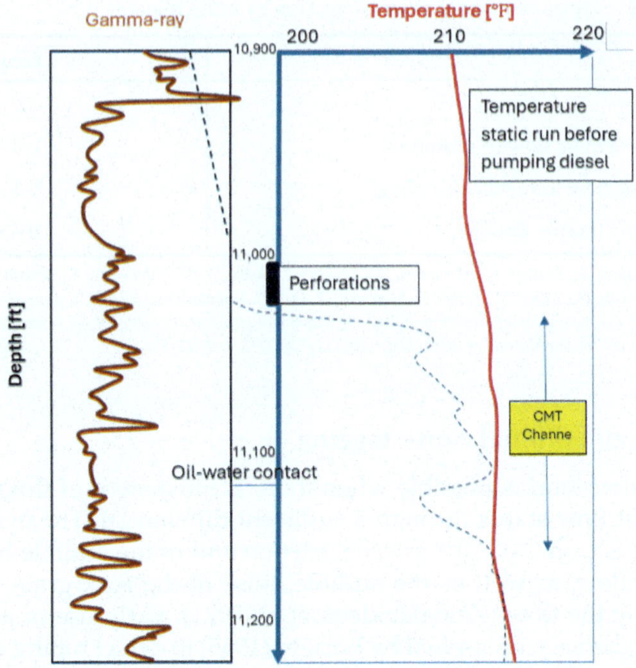

FIGURE 7.3 Temperature composite profile log before cement squeeze. Temperature passes showing change in profile post injecting cooler fluid into perforations (Allouche et al., 2006).

or static, injecting, bleeding down, or pressuring up an annulus. The data can then be compared under the different wellbore conditions.

7.2.4 Case studies

7.2.4.1 Literature case study

After a well has been completed, temperature logs can help assess whether channeling behind the casing is occurring. Fig. 7.3 provides a typical example of temperature log showing the change in temperature profile between runs. The first run (the red solid curve) is the static temperature conditions, although there is slight temperature movement in the data. The second run (black dashed curve) was run a short time after pumping and injecting 80 bbl of diesel into the perforations, showing a large temperature decrease above the perforations and temperature variation down to the oil/water contact and even lower. However, for a well-cemented and integrated well, temperature drop is not expected because the cool fluid can be in touch with zones under the perforations. This observation can indicate hydraulic communication behind the casing. In such cases, it might be needed to carry out remedial cementing to seal the annulus and reduce the water influx (Allouche et al., 2006). This log enables the operating company to evaluate the cement bond and select depths to perforate for cement squeezing to shut off water channeling zone that was impacting production. Perforation should only be performed in well-cemented intervals.

7.2.4.2 Case study–Temperature and noise logs in a production oilwell

7.2.4.2.1 Background

A free-flow onshore oil well, in a tectonically active region in the Middle East, was drilled in 1977 with the following casing/liner pipes run and cemented: 18 5/8″, 13 3/8″, and 9 5/8″ casing strings, as well as 7″ liner and 5″ liner. Perforations were conducted in the 5″ liner across the fractured carbonate reservoir. A simple 2 7/8″ tubing string was run and hung in the wellbore with no packer. The oil well was produced via the A-annulus (2 7/8″ × 9 5/8″) simultaneously. During production, a downhole problem occurred and completely shut off the well's flow capability; the problem was believed to be because of casing and wellbore collapse. Fig. 7.4A shows the original production schematic.

In 2022 a workover rig operation tagged a collapsed casing obstruction at 3365 m (11,040 ft), which could not be cleared. The decision was made to sidetrack the well in the 7″ casing at 3030 m (9941 ft) KOP. A new 5″ liner was run, followed by 3 1/2″ liner. However, circulating within the 9 5/8″, the rig suddenly noticed a high pressure in the B-annulus (9 5/8″ × 13 3/8″). Note that the sidetracked wellbore was unperforated.

A retrieve test treat squeeze packer assembly was run in the hole, mid-wellbore depth, and determined the leak was in the upper section of the well following leak detection. A decision was made to run a noise and temperature log (Spectral Noise Log (SNL)–High-Precision Temperature (HPT) tool string). A cement plug was placed at 2150 m (7054 ft), and the log was run in the upper section while pumping into the wellbore (down the 9 5/8″ casing). Fig. 7.4B shows the sidetracked well schematic with B-annulus pressure.

The SNL–HTP output is shown in Fig. 7.5, as well as the zoomed-in section from 0 to100 m (33 ft). The temperature derivative curve (DT/DZ) in the fifth track shows that significant temperature changes were detected from the surface to 25 m (82 ft) which indicated fluid flow into and behind the 9-5/8″ casing. The remainder of the logging interval shows no other obvious temperature anomaly.

The log provides insight into dynamic well conditions and potential well integrity damage or failure mechanisms.

The noise volume is shown as a rainbow color palette: with red color standing for high volumes, blue color standing for low volumes, and white color standing for no noise (or noise below tool's threshold).

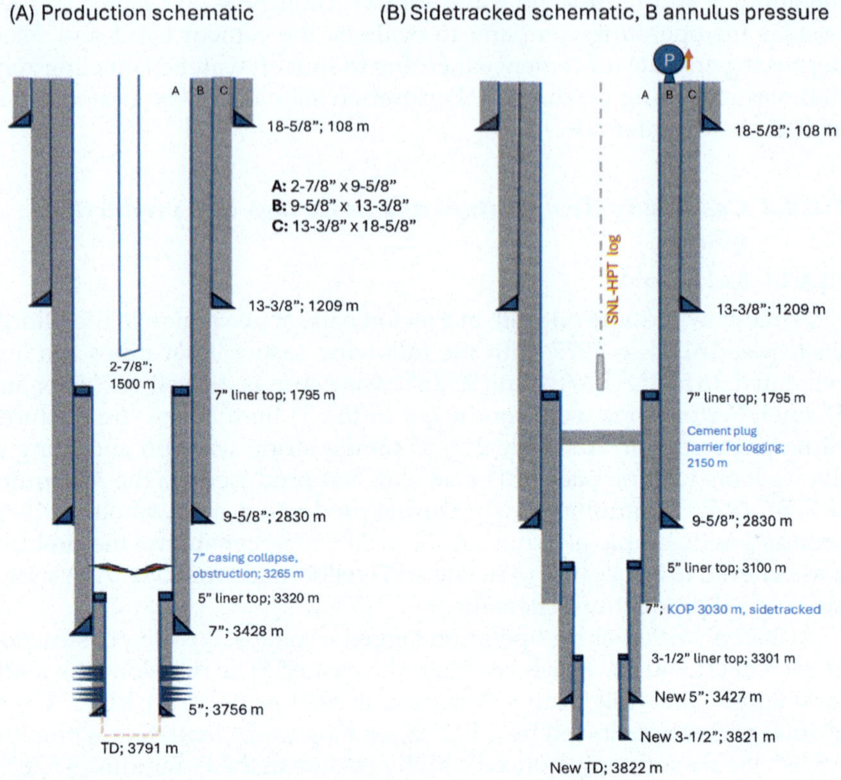

FIGURE 7.4 Well schematics (A) initially producing well and (B) the sidetracked and suspended for logging leak path.

The SNL noise data indicated the following:

- The 1.3 kHz low-volume noise was observed from 10 m (33 ft) to the surface. This is associated with leakage and channeling.
- A very low volume noise (<0.7 kHz) was observed from 90 m (295 ft) to 10 m (33 ft), which is associated with borehole flow.
- The SNL tool did not record any noise, even with very low frequency at depths below 90 m (295 ft), which confirms the lack of any fluid movement within or behind the wellbore from 90 m to the bottom of the logged interval.

The lack of any noise recorded below 90 m (295 ft) as well, along with the noise and temperature response above 25 m (82 ft), suggest the possibility of having a 9 5/8″ casing failure (burst, collapse, or corrosion, or all). The separation between the temperature logs recorded by HPT and well temperature (WTEP) tools indicated the integrity issue to be around this

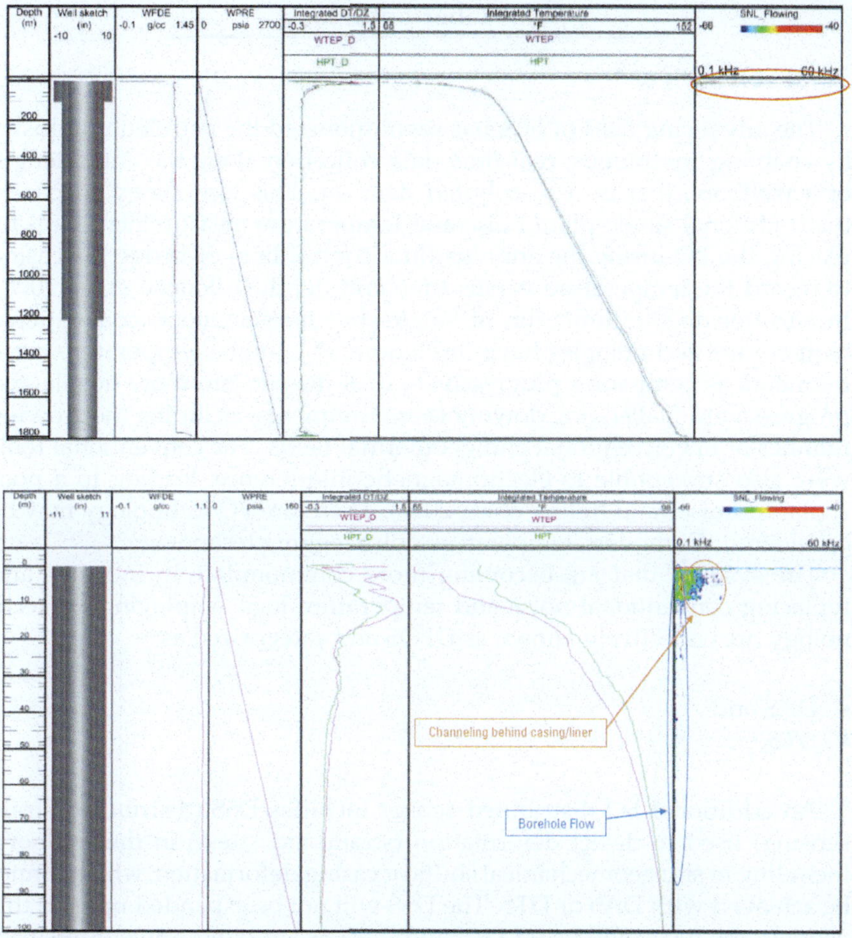

FIGURE 7.5 The spectral noise log (SNL)–HPT (noise temperature) log of the well while injecting water; top (the entire log interval and bottom) log zoomed 0–100 m.

depth; this possibility could be further analyzed by running a corrosion log (e.g., Electromagnetic Defectoscope System (EMDS) tools).

7.2.4.2.2 Outcome

The cellar was dug down, and the upper section of the casing strings was cut, cleaning up the cement behind them up to a depth of 8 m (26 ft). A new joint was welded with a remedial cement job reinstating barriers. The noise and temperature log data enabled the client to recognize the failure point, remediate accordingly, and proceed with successfully perforating and putting the well back into production.

7.3 Fiber optic logs

7.3.1 Overview

The advancing field of FOs has revolutionized the petroleum industry by enabling continuous, real-time data collection along the entire length of a wellbore, that is, a distributed data set. This technology surpasses the traditional single-point noise and temperature measurement tools by placing the FO inside the annulus (in a control line) or inside the casing to record the temperature versus time and depth (Allouche et al., 2006). In addition to the limitation of single-shot measurement, conventional temperature technologies have limitations in extreme temperature environments as companies progressively drill deeper into more hostile and progressively challenging downhole environments, including the growing number of unconventional and geothermal fields. The conventional tools were also susceptible to electromagnetic interference, leading to a poor signal-to-noise ratio (Johny et al., 2021). Therefore FO technology may be considered the modern temperature noise logging technology.

The systems that are becoming more commonplace in industry and replacing conventional noise and temperature logs, employing FO technology, are collectively known as DFOS and categorized as:

- DAS and
- DTS.

An additional FO distributed sensor includes DSS (distributed strain sensing) used to detect deformation (strain and stress) in the wellbore, including in situ geomechanical shifts or casing deformation, which cannot be achieved with DAS or DTS. The DSS will not be expanded upon in this chapter; the focus will be on DTS and DAS as a modern cement integrity and fluid movement detection technique.

The DFOS can be utilized within the upstream oil and gas industry to evaluate wellbore integrity leak detection; casing and cement integrity; production allocation and optimization; diagnosing hydraulic fracture efficiency, geometry, and propagation; determining TOC; or monitoring CCUS projects and the affiliated reservoir behavior as the injection places strain and pressure changes, which may induce a fault activation or caprock breach (Roux et al. 2022), to name but a few of the applications. Jin et al. (2019) mentioned using the DTS to measure borehole temperatures during steady states and using DAS to measure borehole flow velocities by tracking temperature slugging signals during transient states. Fig. 7.6 shows downhole applications. It can also be used at surface for seismic acquisition or to detect pipeline leaks, as well as being applied in other industries.

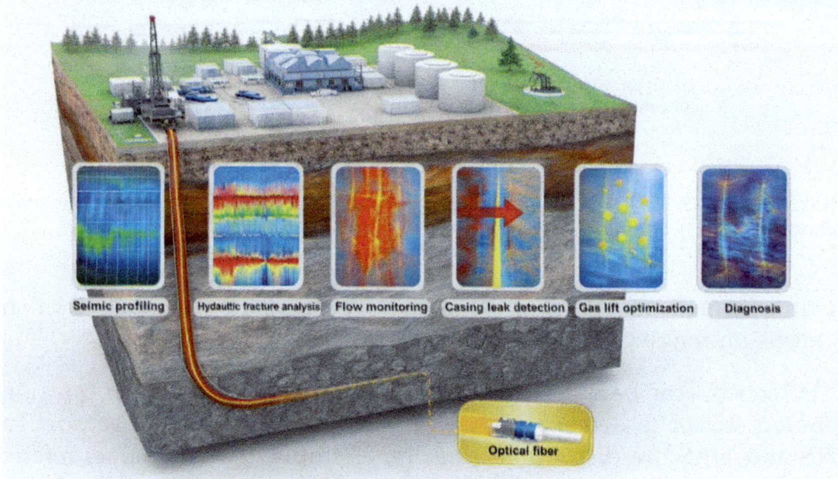

FIGURE 7.6 Example of downhole wellbore applications of a DFOS system. *(Redrawn from Ashry, I., Mao, Y., Wang, B., Hveding, F., Bukhamseen, A., Ng, T. K., & Ooi, B. (2021). A review of distributed fiber–optic sensing in the oil and gas industry. Journal of Lightwave Technology, 40, 1407–1431. https://doi.org/10.1109/JLT.2021.3135653).*

FOs were initially deployed as a permanent installation in the early 1990s (Lovell, 2013) and have since continued to advance technologically, becoming increasingly valuable within the industry as an effective wellbore diagnostic method.

The basic elements that make up the DFOS system include:

a. The **optical fiber**, which itself is the sensor. The most common material is silica (as glass), which has a high heat resistance, can transmit light, and is durable in the oil and gas environment. The optic filament may then be encapsulated in a cladding such as a cable or carbon rod. A jacket or light-blocking coating surrounds the cable. The optical fiber is simultaneously the sensor and the data transmission line.

b. The **laser plasma light source** sends a pulse of light to the optical fiber, which travels the entire length of the fiber. As the light traverses the fiber it encounters glass particles, causing scattering, which is then captured back at the surface, referred to as "backscatter." The time it takes for the backscattered light to reach surface delineates the distance along the fiber sensor (Khankishiyev et al., 2024).

c. The **receiving interrogator** at surface, that is, the acquisition unit, receives the backscatter signal, which travels back via the optical fiber, which contains the data of the acoustic and/or temperature response.

d. **Data analytics software** to produce final output of the processed data in a visualized format for the client. Data may also be monitored in real

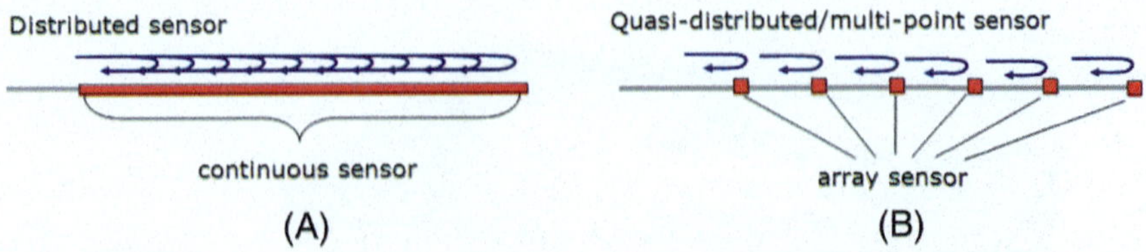

FIGURE 7.7 (A) Distributed fiber-optic sensor and (B) quasidistributed fiber-optic sensor (Johny et al., 2021).

time by analysts via satellite where fast response is required for on-site decision-making.

FO sensors can be installed as a continuous distributed sensor, quasidistributed sensor (a dense array of sensors), or a single-point sensor. The DTS and DAS are distributed sensors, as implied in the name; refer to Fig. 7.7.

A small percentage of the scattered refracted light within the fiber travels back to the origin and is detected at the surface. Without delving into the details of the technology, at a very basic level, the DAS system is a measurement of the Rayleigh backscatter effect occurring at the same wavelength as the transmitted light signal. A DAS system, at extremely low frequency band (<0.1 Hz), is highly sensitive to small temperature variations as low as 10^{-5}°F (Jin et al., 2019).

The DTS system is a measurement of the Raman backscatter effect, which provides information on temperature changes. Raman backscattered light consists of two bands: Stokes and anti-Stokes. The intensity of the anti-Stokes band changes with temperature, and the other stays constant. The temperature fluctuations are based on the ratio of the anti-Stokes to Stokes intensity.

Brillouin scattering is mostly used in DSS sensing and is not covered in this chapter.

Refer to Fig. 7.8 highlighting a wavelength versus scattered intensity plot indicating the Rayleigh, Raman, and Brillouin responses.

Deployment methods for DFOS systems include:

a. Permanent for ongoing well monitoring. Often strapped to the exterior of tubing/casing (see Fig. 7.9A and B).
b. Conveyed in the wellbore, for example, via coil tubing or wireline (C) and (E)
c. Deployment, by pumping downhole (D) in the same figure, or
d. Utilizing gravity to drop a "dart" or "torpedo," sometimes referred to as the disposable method, as the fiber and encasing are not retrieved but disintegrate with time (see Fig. 7.10). This method is economic, requires a low footprint and minimal personnel, and is simplistic to set

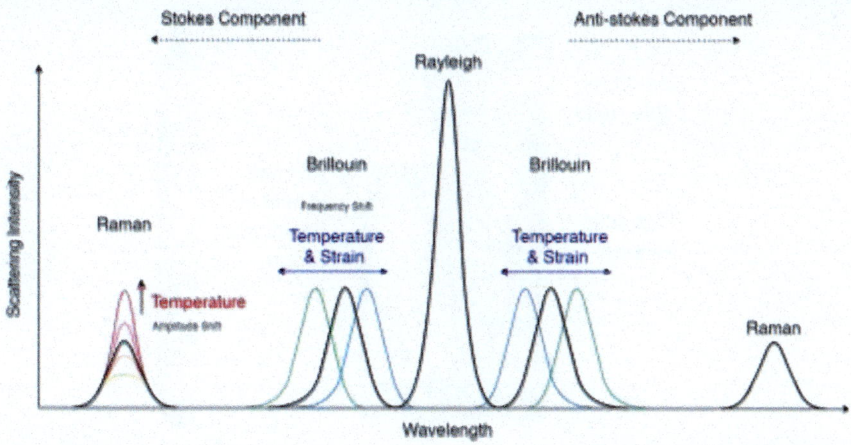

FIGURE 7.8 Rayleigh, Raman, and Brillouin scattering intensity in optical fibers (Anjana et al., 2024).

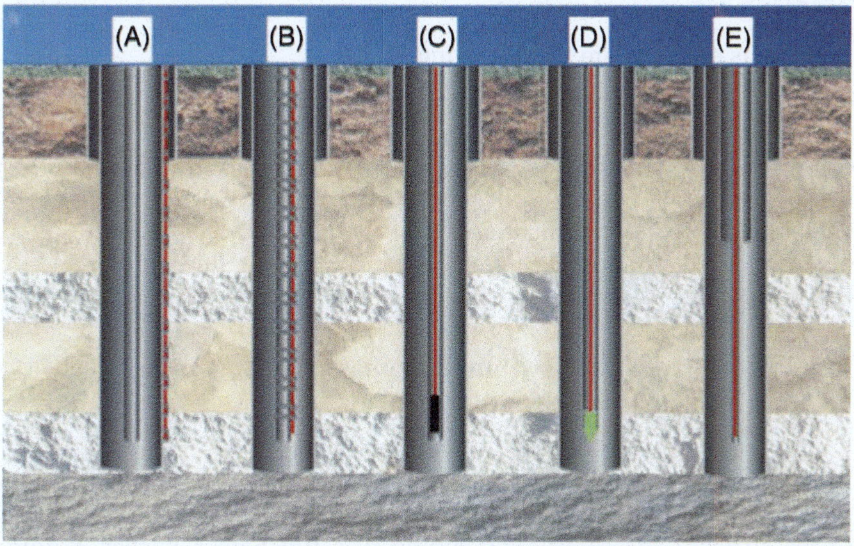

FIGURE 7.9 Fiber optic (*red*) deployment methods: (A) outside of casing, (B) outside of tubing, (C) via wireline, (D) pumped downhole or uses gravity, free hanging, and (E) coil tubing.

up and operate. This system is ideal for remote regions with limited resources and difficult access; residential areas where background noise may impact the data, or larger equipment would affect the residents adversely; and offshore locations where larger units such as coil tubing,

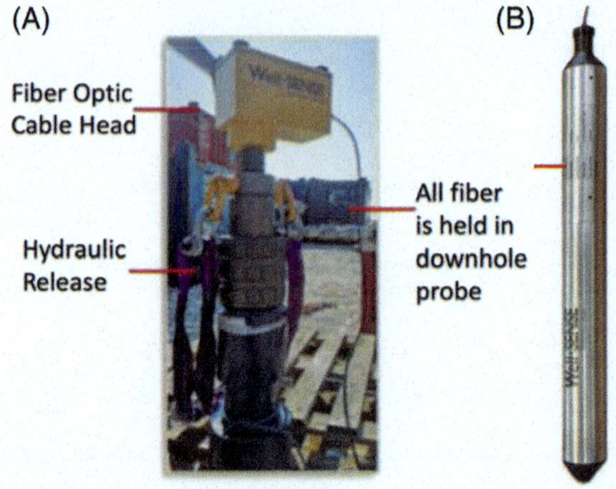

FIGURE 7.10 (A) Disposable fiber deployment system and (B) the "torpedo" containing the fiber optic cable (Courtesy Well Sense).

wireline, or rigs may not be readily available. There is no BOP required for the deployment method.

Benefits of DFOS are the following:

- High-precision data,
- Real-time data for speed of optimizing decision-making,
- Fast data transmission and high band-width capacity (Jinke et al. 2005),
- Cost-effective options,
- Reliable with low maintenance,
- Corrosion resistance, inert to oil and gas wellbore fluids, and
- Multiple installation options, permanent or temporary.

7.3.2 Case study – Distributed acoustic sensing on hot well with sustained casing pressure

Data courtesy Well Sense (www.wellsense.co.uk)

A remote, deep, hot onshore well was drilled with surface casing, intermediate casing, and production casing, with each string cemented to the surface. The completion tubing was run with a packer. Bottom-hole temperature was more than 200°C, leading to cementing challenges with exposure to sour gas. The production casing underwent multiple high-pressure cycles. The single packer failed, and an attempt was made to cement the inner tubing in place above the packer by perforating and circulating cement into the A-annulus (tubing × production casing).

The well surface pressure monitoring indicated pressure on the tubing, as well as annuli A, B, and C. Tubing and A-annulus pressures were of similar value. The B-annulus's pressure was greater than the A-annulus' pressure.

An investigation made by bleeding down the surface pressures established that the tubing and A-annulus were in direct hydraulic communication from an unknown location, and the gas leak path into A-annulus or tubing from below was unknown. The B-annulus bled down and built back up, with no pressure change on the A or C annuli. The C-annulus bled down and built back up with no change on A- or B-annulus.

The well integrity testing processes had established there were no leaks via wellhead seals between adjacent annuli, confirming the problems were originating downhole and not related to the surface equipment leak paths.

There were multiple positions on the production casing where damage had been caused by multifinger caliper. During drilling, it was reported that of 21 joints of run intermediate casing were replaced because of crimping. It was uncertain whether there were leaks via cement, or casings, or both.

The well was planned for abandonment, but the leak paths required determination prior to designing the abandonment to ensure regulatory and company standards were met. The decision was made to run the DFOS in a remote location.

The Well Sense FLI (Fiber Line Intervention) deployable system was selected. The product was a glass FO core with cladding suitable for extreme downhole conditions. The FO cable was stored in a probe that was dropped from the surface and unravels as it descends in the wellbore to the required depth. The probe body was made from aluminum in this case. At the cessation of the data gathering the glass fiber, which is very strong in tension but very weak in shear, is cut at surface, and as the cable falls in the wellbore, it breaks into very small fragments and falls to the bottom. The volume of debris remaining is minor because of the minute diameter of the optical fiber. The aluminum probe will degrade with time and not cause any intervention problems. For production scenarios the FLI probe would be made from a water-soluble magnesium alloy instead. The DAS results are presented Figs. 7.11 and 7.12.

7.3.3 Interpretation of results

By utilizing the Well Sense's FLI deployable technology with DAS/DTS sensing systems the client was able to establish the following to allow for plug and abandonment planning:

a. The C-annulus (surface × intermediate casing) pressure was fed from a formation below the surface casing shoe. There was slow gas migration via the intermediate casing cement.

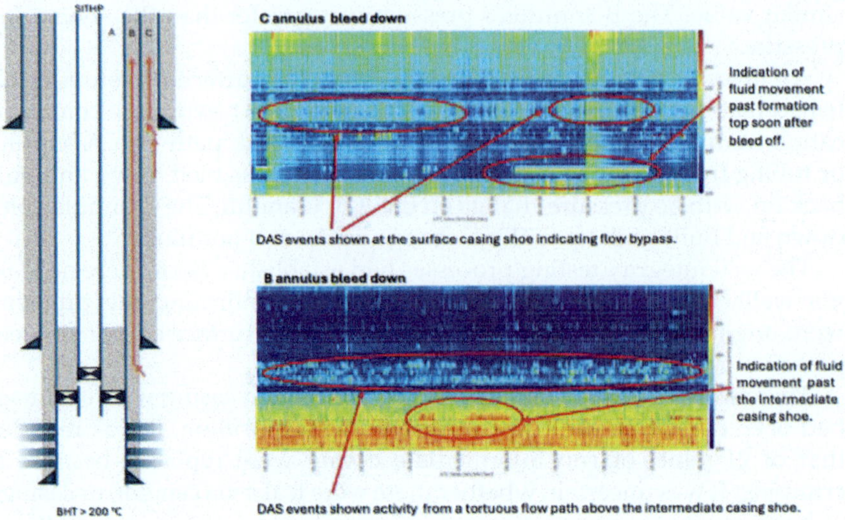

FIGURE 7.11 Distributed acoustic sensing (DAS) results of leak paths when bleeding down C- and B-annulus, respectively (Courtesy Well Sense).

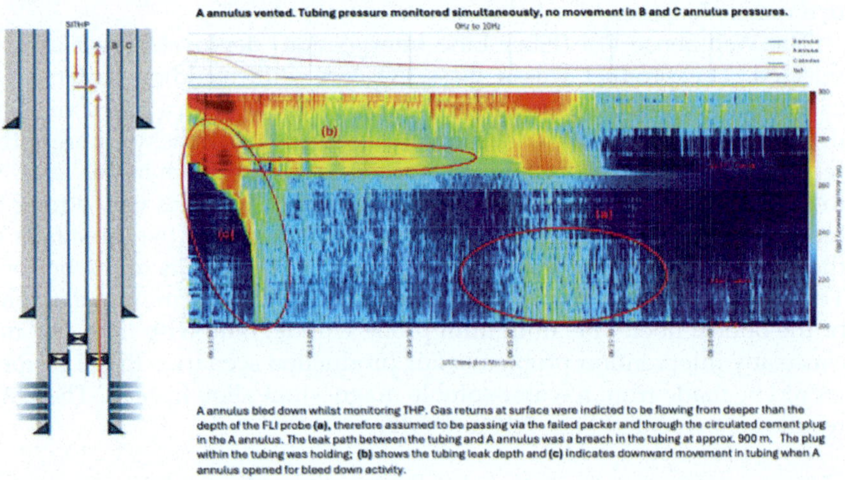

FIGURE 7.12 Distributed acoustic sensing (DAS) results of leak paths into A-annulus from below the packer and between tubing and A-annulus (Courtesy of Well Sense).

b. The B-annulus (intermediate × production casing) pressure was travelling from below the intermediate casing shoe. There was slow gas migration via the production casing cement.
c. The A-annulus (production × tubing string) pressure was being fed from a deeper depth than the FLI bottom probe (run to above the

internal tubing plug). Prior to cementing in the A-annulus the packer had failed. The most credible scenario is the A-annulus gas pressure bypassing the failed packer and migrating through the cement in the A-annulus.

d. The dominant communication between the tubing and the A-annulus was found to be higher up in the wellbore. The plug in the lower section of the tubing, above the cement circulation ports, was not leaking.

Any leak paths below the probe and packer depth are not able to be determined; however, in this instance, it was not required, as there was no intention to remove any wellbore equipment below the TOC within the A-annulus.

The diagnostic information enabled abandonment planning to proceed with known variables, thus increasing efficiencies and certainties in the equipment selection and operational approach to be taken.

7.4 Summary

It may necessitate undertaking additional evaluation methods over and above typical cement bond log data to provide confidence in the cement barrier validation and determine possible fluid movement breaching regulations, standards, and policies. Dated technologies, such as radioactive tracers, provide the minimum information. Acoustic, sonic, and ultrasonic cement bond logs acquire data to varying degrees of confidence for cement-to-casing and cement-to-formation bond quality; however, they do not quantify or negate fluid or gas movement behind pipe. Therefore conventional temperature and noise logs have expanded the depth of wellbore integrity understanding, but with their single-point-in-time data.

The single-point limitation has been overcome using FO technology. Therefore FO logs, particularly DFOS, are increasingly gaining a reputation as a valued tool for providing real-time continuous analysis, useful for cement bond integrity by providing real-time, continuous monitoring of parameters such as temperature and strain along the entire length of the well. These systems, which can be permanent, semipermanent, or temporary, offer a detailed view of the well's integrity. By detecting anomalies such as microannuli or areas of weak cement bonding, FO logs help identify potential cementing issues that may compromise well integrity. This technology offers a noninvasive and efficient method for assessing cement placement and identifying zones requiring remedial attention.

Acknowledgments

The author wishes to cordially thank Mr. Rob Crawford (at Well Sense) for providing information on FLI product and DFOS system as well as the hot well case study.

References

Allouche, M., Guillot, D., Hayman, A. J., Butsch, R. J., & Morris, C. W. (2006). Cement job evaluation. In E. B. Nelson, & D. Guillot (Eds.), *Well cementing* (2nd ed., Chapter 15). Schlumberger.

Al-Sabea, S., Patra, M. K., Abu-Eida, A., Dhafiri, J. M., Gobran, M. H., Al Rasheed, F. A., & Prosvirkin, S. (2022). Defining the zonal contribution in a horizontal well with cased hole ICD's completion using innovative integration of multi arrays production logging, spectral noise and high resolution temperature log. In *Paper presented at the Offshore Technology Conference Asia, Virtual and Kuala Lumpur* March 2022 https://doi.org/10.4043/31593-MS.

Anjana, K., Herath, M., & Epaarachchi, J. (2024). Optical fibre sensors for geohazard monitoring – A review. *Measurement, 235*. https://doi.org/10.1016/j.measurement.2024.114846.

Ashena, R., Thonhauser, G., & Dianati, D. (2014). Assessment of reliability of cement bond evaluation with some interesting field case studies. In *Paper presented at the Abu Dhabi International Petroleum Exhibition and Conference* https://doi.org/10.2118/172039-MS.

Ashry, I., Mao, Y., Wang, B., Hveding, F., Bukhamseen, A., Ng, T. K., & Ooi, B. (2021). A review of distributed fiber–optic sensing in the oil and gas industry. *Journal of Lightwave Technology, 40*, 1407–1431. doi:10.1109/JLT.2021.3135653.

Enright, R.J. (1955). Sleuth for Down-Hole Leaks. Oil & Gas Journal, Vol. 53, No. 43, 28 February 1955, pp. 78–79.

Fertl, W. H., Pilkington, P. E., & Scott, J. B. (1974). A look at cement bond logs. *Journal of Petroleum Technology, 26*(1974), 607–617. https://doi.org/10.2118/4512-PA.

Ghalem, S., Draoui, E., Mohamed, A., Keshta, O., Serry, A. M., Al-Felasi, A., Berrim, A., Chaker, A., Hicham, F., Maxim, G., Alfiya, A., Irina, & Aslanyan, A. (2012). Innovative noise and high-precision temperature logging tool for diagnosing complex well problems. In *Paper presented at the Abu Dhabi International Petroleum Conference and Exhibition* November 2012 https://doi.org/10.2118/161712-MS .

Harris, K. (2021). Chapter eleven: Cement job evaluation. *Applied well cementing engineering* (pp. 523–567). Gulf Professional Publishing ISBN 9780128219560. https://doi.org/10.1016/B978-0-12-821956-0.00010-9 .

Jin, G. E., Friehauf, K., Roy, B., Constantine, J. J., Swan, H. W., Krueger, K. R., & Raterman, K. T. (2019). Fiber optic sensing-based production logging methods for low-rate oil producers. In *Paper presented at the SPE/AAPG/SEG Unconventional Resources Technology Conference* July 2019 https://doi.org/10.15530/urtec-2019-943 .

Jinke, H., Wenlong, L., Yubao, L., & Yugang, T. (2005). Application of fibre downhole video logging tool in administering of casing damage well WLT. *Petroleum Instruments, 4*, 007.

Johny, J., Amos, S., & Prabhu, . R. (2021). Optical fibre-based sensors for oil and gas applications. *Sensors, 21*(18), 6047. https://doi.org/10.3390/s21186047.

Jutten, J., & Morriss, S. L. (1990). Chapter 16: Cement job evaluation. *Well cementing. Developments in petroleum science*. Elsevier https://doi.org/10.1016/S0376-7361(09)70314-0.

Khankishiyev, O., Salehi, S., Hasanov, G., & Hu, Z. (2024). Application of distributed temperature sensing (DTS) in geothermal wells. In *Proceedings, 49th workshop on geothermal reservoir engineering*. Stanford University https://pangea.stanford.edu/ERE/db/GeoConf/papers/SGW/2024/Khankishiyev2.pdf.

Lee, Y., & Lee, Y. S. (2019). A new submersion detection sensor using two resistance temperature detectors operating on the thermal equilibrium principle. *Sensors, 19*(19), 4310. https://doi.org/10.3390/s19194310.

McKinley, R. M., Bower, F. M., & Rumble, R. C. (1973). The structure and interpretation of noise from flow behind cemented casing. *Journal of Petroleum Technology, 25*(1973), 329–338. https://doi.org/10.2118/3999-PA.

McKinley, R. M., & Thornhill, J. T. (1994). *Temperature, radioactive tracer, and noise logging for injection well integrity*. Robert S. Kerr Environmental Research Laboratory, Office of Research and Development, US Environmental Protection Agency.

Pilkington, PE. (1992). Cement evaluation – Past, present, and future. *Journal of Petroleum Technology*, *44*(1992), 132–140. https://doi.org/10.2118/20314-PA.

Roux, P.-F., Patel, A., Johnston, W., Bale, D., & Nazarenko, P. (2022). On advantages of using distributed fiber optic sensing for monitoring in CCUS (October 26, 2022). In *Proceedings of the 16th Greenhouse Gas Control Technologies Conference (GHGT-16) 23-24 Oct 2022* https://ssrn.com/abstract=4283716 or. doi:10.2139/ssrn.4283716.

"Proceeding">Sadeghi, A., Kord, S., Bahari Moghaddam, M., & Salehi, R. (2024). Well integrity evaluation by integrating noise and temperature logs. In 85th EAGE annual conference & exhibition – Workshop programme. 10.3997/2214-4609.202410455.

Sharma, J., Adnyana, G., Barnes, D., Mims, D., & Behrens, R. (2021). Temperature logging guidelines and factors that affect measurement accuracy in steamfloods. *Journal of Petroleum Science and Engineering, 196*, 107727. https://doi.org/10.1016/j.petrol.2020.107727.

Zeinalabideen, M. J., Al-Hilali, M. M., & Savinkov, A. (2021). State of the art of advanced spectral noise and high-precision temperature logging technology utilization in Iraqi oil fields: An integration approach to diagnose wells performance complications. *Journal of Petroleum Exploration and Production Technology, 11*, 1597–1607. https://doi.org/10.1007/s13202-021-01152-y.

CHAPTER

8

Corrosion and other logs

Rahman Ashena[a], *Zanko Asadi*[b]

[a]Smart Energy Data Analytics (SEDA)-Group, Houston, TX, United States, [b]Independent Consultant

8.1 Introduction

When the cement is debonded from the casing, it creates a pathway for fluid migration. If these fluids are corrosive, they may lead to corrosion of the casing. Conversely, it can be stated that when we detect corrosion on the outer side of a casing, it signals poor bonding or debonding of cement from casing as well as existence of a pathway for fluids bypassing the cement sheath or fluid migration. Corrosion is especially problematic in high-pressure or sour environments, where corrosive substances, such as CO_2 or H_2S, may increase corrosion rates. Corrosion logs play a crucial role in identifying when casing corrosion has advanced to the point of compromising the cement sheath, allowing for prompt remediation. As a result, they can effectively complement other methods of cement integrity evaluation.

Corrosion can occur either inside the casing, where it interfaces with drilling or completion fluid, or outside the casing, at the junction with the cement sheath. In some cases, corrosion may affect both areas simultaneously.

- If there is fluid or gas channeling behind the casing because of poor or inadequate cement bonding, erosion and corrosion can develop, especially if formation fluids contain CO_2 or H_2S. In these cases, detecting corrosion behind the casing becomes challenging, as there is no direct method to observe or measure it. The industry has, however, developed methods to deal with the challenge.
- Corrosion inside the casing typically results from the choice of a through-annulus completion or production scenario, along with the improper selection of completion fluid. In some older wells, production

Well Cement Integrity
DOI: https://doi.org/10.1016/B978-0-443-51564-4.00008-4

may still occur through the annulus between the completion string and the production casing (Ashena et al., 2021), and such wells often face significant corrosion challenges. This can present a major issue for the completion packer, potentially causing leakage. If these wells are recompleted or undergo workover operations, there will not only be a significant challenge in setting the retrieve test treat squeeze packers (used for workover) but also with the downhole pumps.

Corrosion logs can help identify signs of corrosion both inside and behind the casing, utilizing techniques that contribute to assessing the overall casing condition and the well's integrity status. Techniques used to take corrosion logs are usually based on either electromagnetic (EM) logging tools or ultrasonic methods. EM corrosion logging tools estimate (average) metal thickness and metal loss percentage for each individual casing pipe. Because of the ease of data acquisition through logging operations and the ability to model remaining metal thickness per barrier, demand for EM corrosion logs has been sharply growing across the petroleum industry (Macki et al., 2017). There are several corrosion logging tools in the market; an example of the EM method is by a tool called EM defect scope or pipe inspection tool. An example tool for ultrasonic acquisition is the CAST tool (by Halliburton) or USIT (by SLB).

- The EM tools are used for internal and external casing corrosion estimation and analysis. Briefly the tool is built with a strong EM core that sets up current flow in the casing based on Faraday's law for EM induction with time going through transmitter coils. Usually the tool contains two to three vertical coils and can contain transverse coils. Vertical coils are most sensitive to vertical features, as they generate lateral eddy currents (see Fig. 8.1). For this reason, some tool manufacturers use transverse coils to add sensitivity to horizontal features. Several pads mounted on the sonde body of the standard tool investigate the total circumferential area of the casing. Two sets of measuring devices are mounted on each pad: one measures the eddy current, and the other measures the flux leakage. The eddy current signal varies with internal changes in the pipe surface. The flux leakage responds to total defects to the internal and external casing. Responses vary according to pitting size and depth (Wheeler, 1998). The amount of data recorded is very large to accommodate the tool's vertical resolution and the high sampling rate (120 samples/ft). The data is sent to a computing center where the logs are processed, and the results are presented. The software uses calibration to normalize all data responses. The software then uses the responses of both the eddy and flux measurements to determine the type, placement, and percentage of penetration of the corrosion; therefore the corrosion is classified as internal, external, or circumferential.

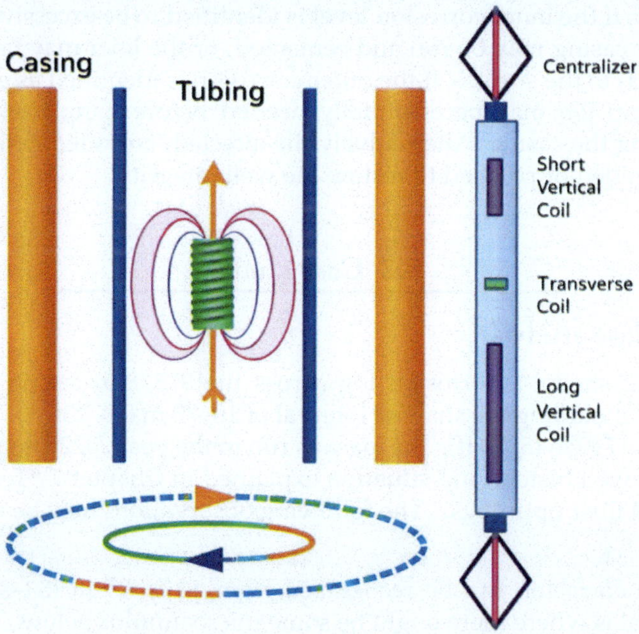

FIGURE 8.1 Electromagnetic (EM) field induction through tool excitation coil and forming longitudinal eddy currents. *(Redrawn from Macki, A. A., Salmi, A. A., Duggan, T., Delgado, E., Kechichian, J., Busaidi, H. A., & Azri, N. A. (2017). Electromagnetic corrosion logs, insights from yard test. In Paper presented at the Abu Dhabi International Petroleum Exhibition & Conference, Abu Dhabi, UAE November 2017. https://doi.org/10.2118/188672-MS).*

- The main physical limitation of the tool is that further out from origin of the source, the signal weakens. The signal reduction over time is close to logarithmic behavior, as it quickly deteriorates moving from one time channel to another. Therefore as the technology aims to investigate deeper barriers, the tradeoff with a lower signal-to-noise ratio becomes a serious consideration (Macki et al., 2017).

In addition, there are two more techniques, which are good for determining inside corrosion. First, caliper logs are run in the hole with corrosion logs to complement the corrosion logs because caliper measures the inside diameter and thus contributes to detecting the inner erosion and corrosion and quantifying their effect. The 40–60 independent caliper arms are conventionally run to determine the remaining wall thickness of the casing pipe. The other complementary method is by using downhole video cameras to provide a visual inspection of the inner casing surface, contributing to indications on corrosion levels. Based on Wheeler (1998) the depth and temperature ratings of video cameras at that time were 12,500 ft (3810 m) and 225°F (107°C), respectively.

Finally, if the inner corrosion level is identified to be excessively high, a narrower casing may be run and cemented, or the liner may be tied back (extended) to the surface. If the outer corrosion is identified as excessively high, a top job may be essentially needed before tying back the liner or running the casing. Alternatively the affected/corroded casing section may be milled or patched to restore the well integrity.

8.2 Case studies

8.2.1 Case study I

Fig. 8.2 shows a corrosion log across the 9 5/8-in casing at several intervals, including the shallow interval of 10–70 m (33–230 ft). The casing shoe was 1790 m (5873 ft). The log was run in the year 2022 and pertains to the same well history and situation explained in Chapter 7 "Temperature noise and fiber optic logs." The following observations may be taken from the log:

- Minor corrosion can be recognized from 10 to 25 m (33–82 ft). This interval is where there might be some microannulus below.
- Moderate corrosion is recognized from 25 to 50 m (82–164 ft) and 1315 and 1325 m (4314–4347 ft).
- Severe corrosion is observed from 1610 to 1620 m (5282–5315 ft), but no flow behind the casing was observed or detected by the other diagnostic logs.

As mentioned in Chapter 7, flow was recognized behind the casing from around 10–25 m (33–82 ft) to the surface. As corrosion level was recognized as low, this means that the cement integrity issue had begun not long before its detection.

8.2.2 Case study II

Fig. 8.3 shows an EM corrosion log across the 9 5/8-in casing of a vertical well from 25 m (82 ft) to the shoe of 2590 m (8498 ft). The log shows low corrosion except at two intervals of each being around 10 m (33 ft) with moderate corrosion levels:

- 1545–1555 m (5069–5102 ft).
- 2280–2290 m (7480–7513 ft).

8.2.3 Case study III

Fig. 8.4 shows an EM corrosion log across the interval of (935–1010 m/3067–3314 ft) in the 9 5/8-in casing with shoe depth of 2645 m

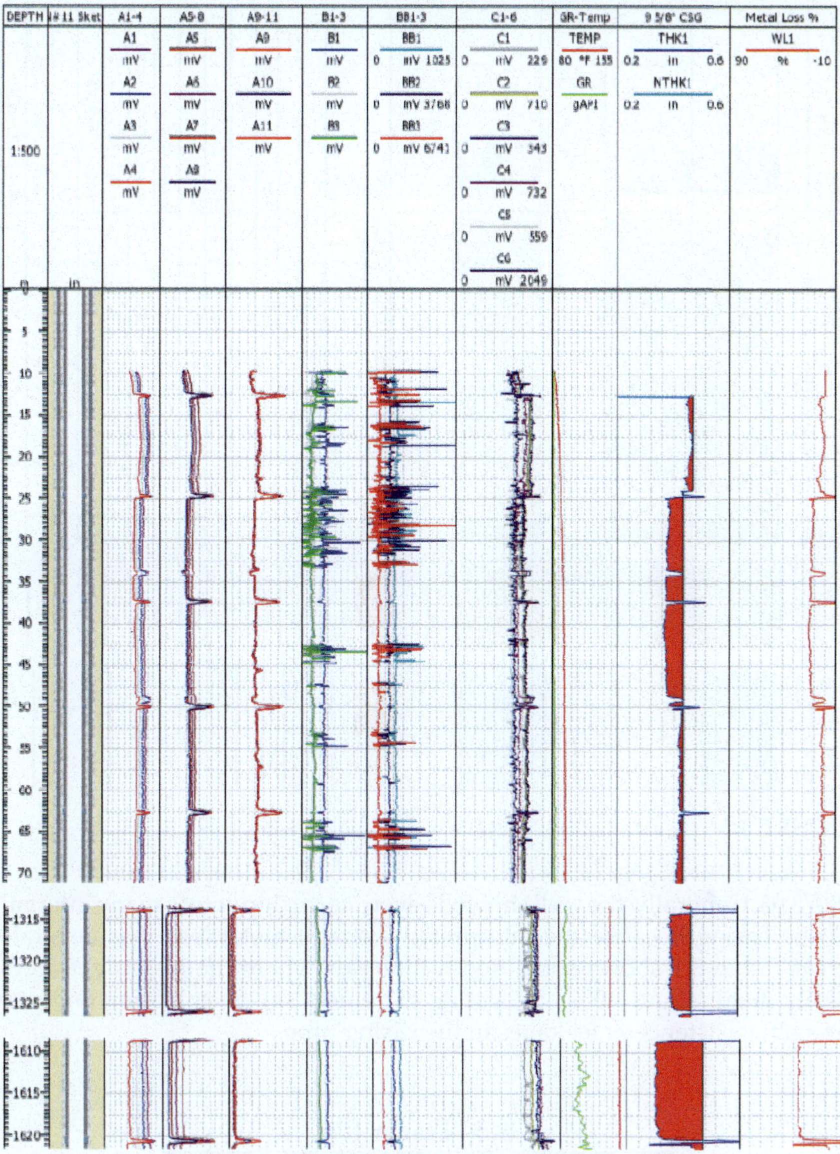

FIGURE 8.2 Case study I with moderate corrosion of casing.

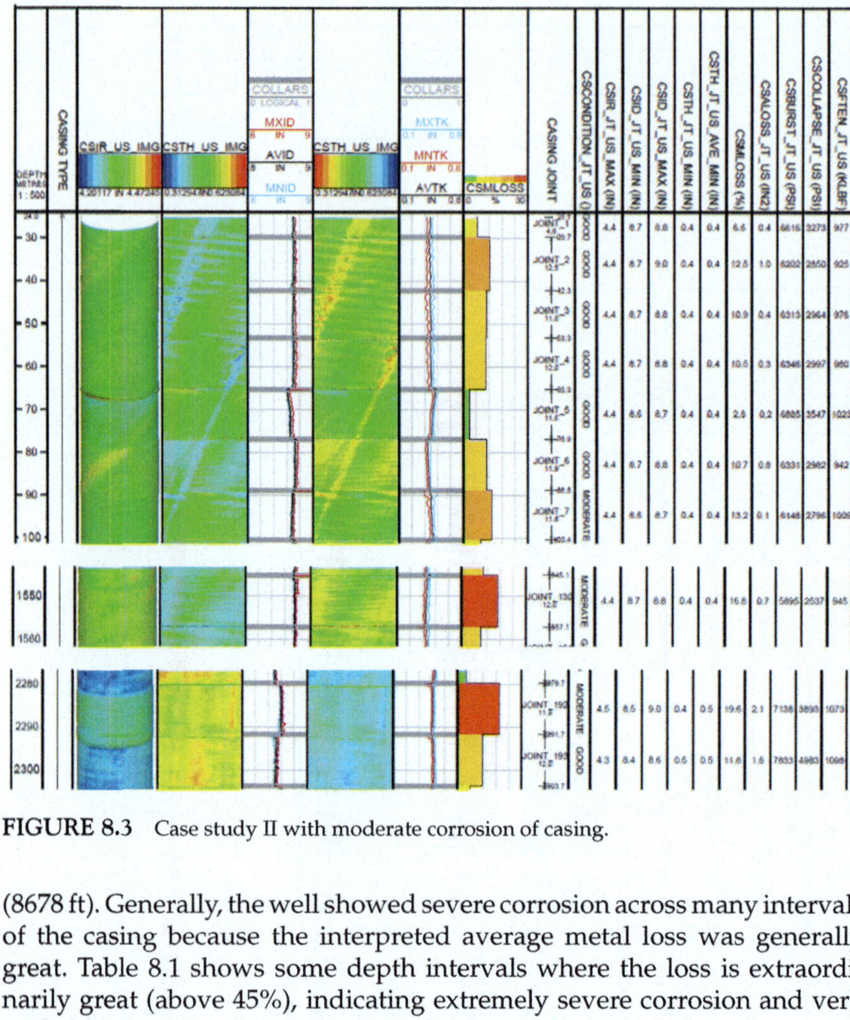

FIGURE 8.3 Case study II with moderate corrosion of casing.

(8678 ft). Generally, the well showed severe corrosion across many intervals of the casing because the interpreted average metal loss was generally great. Table 8.1 shows some depth intervals where the loss is extraordinarily great (above 45%), indicating extremely severe corrosion and very probable existence of leakages in the casing pipe.

8.3 Tracer log

Around the same era as the introduction of temperature logging to find TOC, in the 1930s, it became customary to mix a radioactive tracer, such as carnotite, into the cement slurry. Since then, tracers with a shorter half-life have been utilized as a soluble tracer, iodine-131 (8 days), or as sand or glass beads coated in a radioactive substrate, like iridium-192, with a much longer half-life (74 days). A post cementing logging tool, which includes a gamma-ray sensor, can then detect the change in radiation intensity on

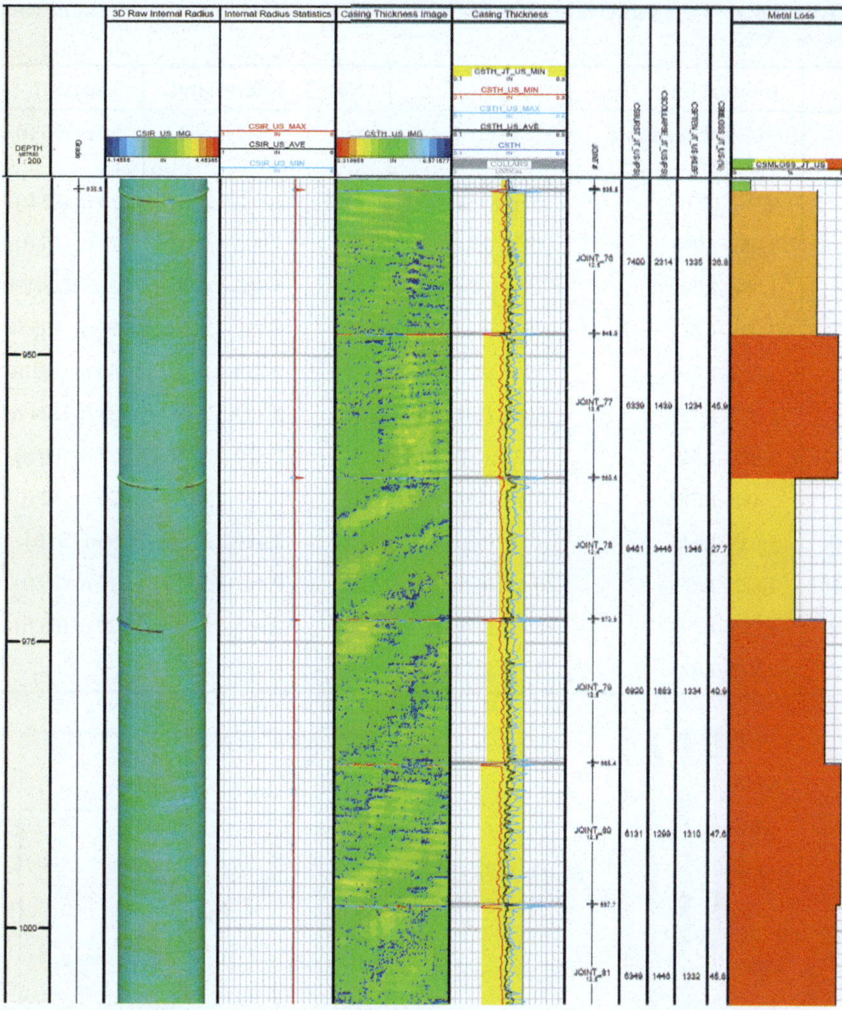

FIGURE 8.4 Case study III with severe corrosion of casing.

the log comparatively to the original gamma-ray log (typically open-hole log data). Running radioactive tracers today for cement evaluation is not a common practice because of material handling health and safety protocols, and conceivably the environmental focus of the impact of the oil and gas industry on ground water and critical aquifers would steer companies away from using this method when there are other options available to evaluate the cement integrity and location in the wellbore.

Fig. 8.5 interestingly shows that by combining temperature log with a water-soluble radioactive tracer, some of the tracer mixes appear in the

TABLE 8.1 Depth intervals across the 9 5/8-in casing for case study III with casing metal loss > 45%.

z	Interval [m]	Length	No	Interval [m]	Length
1	948–960 m (shown in figure)	12 m (39 ft)	14	1,840–1,850	10 m (33 ft)
2	973–1,023	50 m (164 ft)	15	1,860–1,870	10 m (33 ft)
3	1,035–1,048	13 m (43 ft)	16	1,900–1,923	23 m (75 ft)
4	1,060–1,085	25 m (82 ft)	17	1,960–2,075	115 m (377 ft)
5	1,098–1,223	25 m (82 ft)	18	2,135–2,225	90 m (295 ft)
6	1,235–1,260	25 m (82 ft)	19	2,235–2,275	40 m (131 ft)
7	1,273–1,323	50 m (164 ft)	20	2,300-2,338	38 m (124 ft)
8	1,335–1,448	113 m (371 ft)	21	2,350–2,377	27 m (88 ft)
9	1,495–1,610	115 m (377 ft)	22	2,387–2,412	25 m (82 ft)
10	1,635–1,645	10 m (33 ft)	23	2,426–2,435	9 m (30 ft)
11	1,670–1,675	5 m (16 ft)	24	2,492–2,502	10 m (33 ft)
12	1,705–1,790	85 m (279 ft)	25	2,565–2,585	20 m (66 ft)
13	1,815–1,825	10 m (33 ft)			

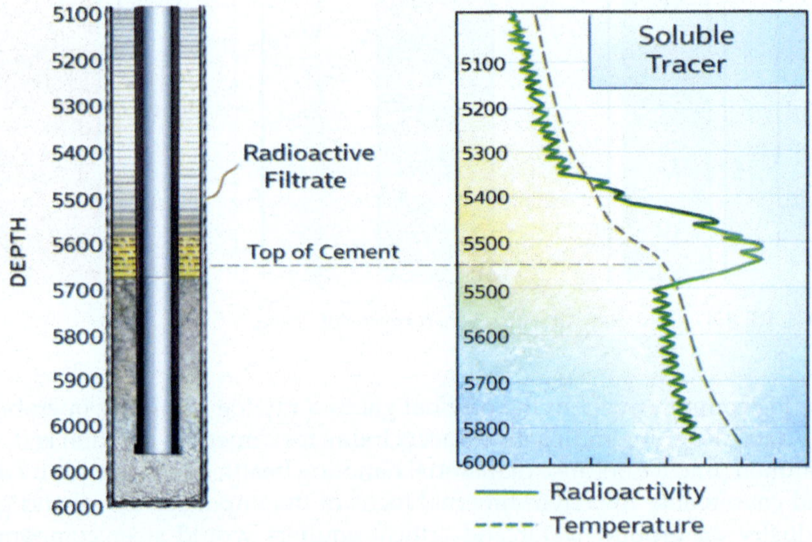

FIGURE 8.5 Radioactive tracer log with temperature log indicating Top of Cement (TOC). *(Redrawn from Gore, G.L. & Terry, L.L. (1956). Radioactive tracer techniques. Journal of Petroleum Technology, 8(9), 12–17).*

cement spacer. The two data curves combined were able to provide a more accurate TOC analysis in this instance.

References

Ashena, R., Bataee, M., Jafarpour, Abbasi, H., Zolotukhin, A., & Mirhashemi, M. (2021). Significant production improvement using optimization of completion and artificial lift: Case studies from South-West Iran. *Journal of Petroleum Exploration and Production Technology, 11*, 359–384. https://doi.org/10.1007/s13202-020-01053-6.

Gore, G.L. & Terry, L.L. (1956). Radioactive tracer techniques. Journal of Petroleum Technology, 8(9), 12–17.

Jin, G., Friehauf, K., Roy, B., Constantine, J. J., Swan, H. W., Krueger, K. R., & Raterman, K. T. (2019). Fiber optic sensing-based production logging methods for low-rate oil producers. In *Paper presented at the SPE/AAPG/SEG Unconventional Resources Technology Conference, Denver, Colorado, USA* July 2019 https://doi.org/10.15530/urtec-2019-943.

Macki, A. A., Salmi, A. A., Duggan, T., Delgado, E., Kechichian, J., Busaidi, H. A., & Azri, N. A. (2017). Electromagnetic corrosion logs, insights from yard test. In *Paper presented at the Abu Dhabi International Petroleum Exhibition & Conference, Abu Dhabi, UAE* November 2017 https://doi.org/10.2118/188672-MS.

Wheeler, R. R. (1998). Improving field economics through the use of modern casing-corrosion logging tools and techniques. In *Paper presented at the SPE Annual Technical Conference and Exhibition, New Orleans, Louisiana* September 1998 https://doi.org/10.2118/49297-MS.

CHAPTER

9

Fundamentals of cement plugs and evaluation

Rahman Ashena[a], *Razak Yakob*[b]

[a] **Smart Energy Data Analytics (SEDA)-Group, Houston, TX, United States,** [b] **Independent Consultant, Kuala Lumpur, Malaysia**

9.1 Introduction

Cement plugs serve as solid barriers within a wellbore, placed to seal off specific sections during various well operations, including drilling, well control, well intervention, and plugging and abandonment (P&A). These plugs play a critical role in preventing fluid migration between zones, mitigating the risk of leaks, and maintaining well integrity.

The process of plug cementing involves placing and setting a cement plug within the wellbore or casing to create a reliable seal, preventing formation fluid from flowing into the wellbore or reaching the surface. The success of cement plugging depends on meticulous planning, proper material selection, and precise placement to ensure an effective and durable seal. According to API (2010) the placement and design of the cement plug should consider well conditions and regulatory requirements. A statistical work by Ashena et al. (2023) in an onshore field in the Middle East shows that 7 out of all 117 kicks (5.98%) occurred during cement plugging.

Cement plugs are set for various reasons, including:

✓ *Controlling lost circulation problems during drilling:*

Mud loss is a drilling problem, which should be cured prior to proceeding to primary cementing. When mud loss cannot be cured by spotting drilling mud with lost circulation materials (LCMs), setting cement plugs may be sought. Setting neat cement slurries (i.e., no additives) are effective for solving small mud losses; however, severe losses require cement mixed with clays, diatomaceous earth, and LCMs (Nelson & Guillot, 2006).

Well Cement Integrity
DOI: https://doi.org/10.1016/B978-0-443-51564-4.00005-9

✓ *Sealing off the lower hole intervals:*
When sidetracking, deviating, or recompleting a well in a shallower zone, the lower section of the wellbore must be isolated. In such cases, kick-off plugs are set to facilitate sidetracking or directional drilling. When sidetracking, deviating, or recompleting a well in a shallower zone, the lower section of the wellbore must be isolated. In such cases, kick-off plugs are set to facilitate sidetracking or directional drilling.

✓ Protecting weak or low-strength formations:
Shallow formations with low strength need to be shielded from high pressures applied during operations such as pressure testing. Cement plugs act as a protective barrier, preventing damage to these vulnerable zones.

✓ *Plug and abandonment (P&A):*
At the end of a well's productive life or for other operational reasons, P&A requires setting cement plugs to permanently isolate the wellbore and prevent any future fluid migration.

9.2 Placement methods

There are several methods to set cement plugs:

9.2.1 Balanced cement plugging

The balanced cement plug method is one of the most common techniques for setting cement plugs in both open- and cased-hole conditions, including plug and abandonment (P&A) operations. In this approach, cement is placed in the wellbore through the annulus of a drill pipe or coiled tubing, leaving a continuous cemented section after the pipe is pulled out of the hole. When coiled tubing is used, a rig is not required, reducing operational complexity and cost. Fig. 9.1 illustrates a schematic of a balanced cement plug, showing the cement slurry in place (left) and the wellbore condition after the drill pipe has been removed. This method is widely applied in various well operations, including:

- Controlling mud loss: Balanced cement plugs are effective for curing partial losses. When mud losses are low (< 10 bbl/hour), a balanced plug may be applied for completely controlling the loss.
- Further well control: In critical cases, especially gas wells or wells with extremely high pressures or H2S content, after killing the well, one or two cement plugs (of say 150 ft/45.7 m thickness) may be installed at a critical, usually at a deep point, for example, near the liner lap (Ashena, 2025). If this action is not made and just mechanical plug and prongs are placed in the completion string, gas might accumulate below the

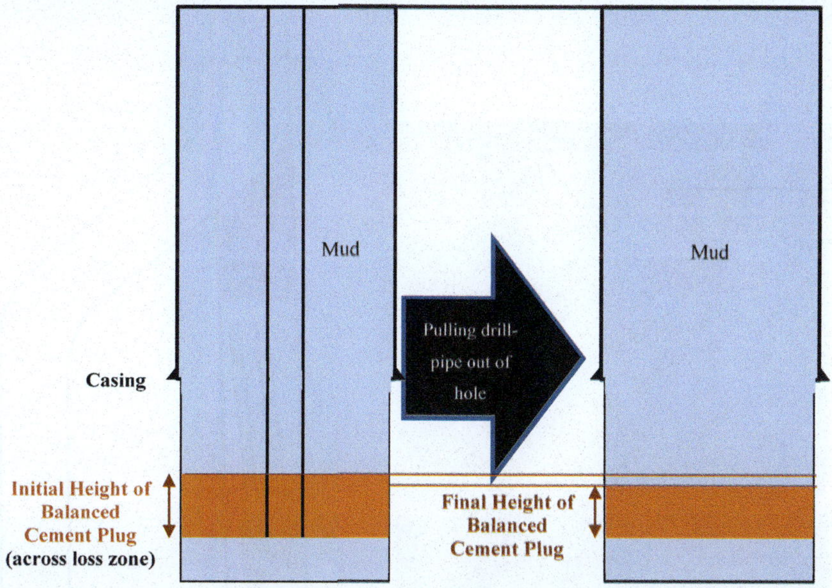

FIGURE 9.1 Schematic of the balanced cement plugging when the cement plug slurry is set (*left*) and after pulling the drill pipe out of the hole.

plug, which would be dangerous when an attempt is made to retrieve the plug.

- Sidetracking: This method is used to drill a secondary wellbore away from the original trajectory. Sidetracking is often necessary for directional drilling (Fig. 9.2A), bypassing an unusable section of the wellbore, or accessing a nearby geological target. In cases where the original wellbore is blocked because of stuck tools, junk, or a collapsed section, a bypass wellbore can be drilled parallel to the original one (Fig. 9.2B).
- Off-bottom completion: In certain open-hole completions the well is drilled to its final total depth (TD), but the casing shoe depth is intentionally set a few feet above TD for specific operational reasons. In such cases the interval between the TD and the casing shoe depth is first cemented using a balanced plug, ensuring proper zonal isolation before running the casing to the bottomhole. Fig. 9.3 illustrates this application.
- Another application of cement plugs is when it is intended to repair the bottom pipe rams. In this case, it is necessary to set one or two cement plugs first. Then the retrievable test-treat-squeeze (RTTS) packer is set at the depth of say 100 ft. Then the bottom pipe rams can be safely disconnected and repaired.

This method remains a preferred solution because of its reliability, versatility, and ability to address various wellbore challenges.

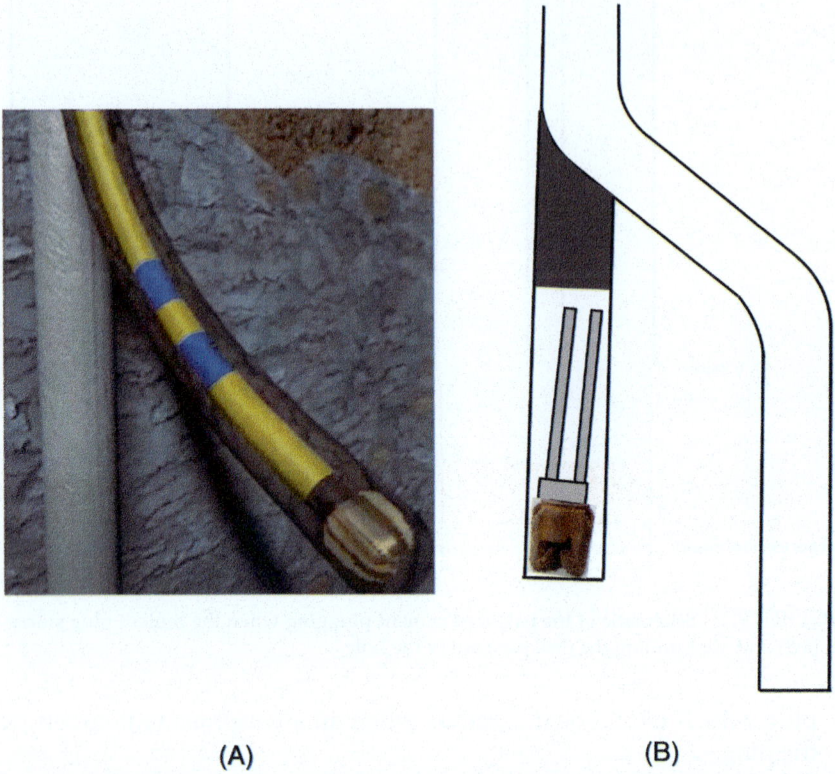

FIGURE 9.2 Balanced cement Plug set at the lower section of the original wellbore for the purpose of directional drilling: (A) kick-off for directional drilling and (B) sidetrack after an unretrievable fish.

9.2.2 Pump and pull

In pump and pull cement plugging method, cement is pumped in the annulus at the same rate as the workstring is pulled out of the hole. Unlike the balanced plug method, this approach does not rely on hydrostatic pressure to maintain the plug in place. This method does not rely on hydrostatic pressure to balance the plug. Table 9.1 lists the advantages and disadvantages of this method, whereas Fig. 9.4 shows the cementing procedure (API, 2021).

Coiled tubing is commonly used in this method because of its ability to facilitate simultaneous fluid pumping and string retrieval with minimal complexity. However, when using drill pipe or jointed tubing, precautions must be taken to prevent U-tubing or fluid free-fall during pipe connections by closing the annular preventer. When drill pipe or tubing (jointed

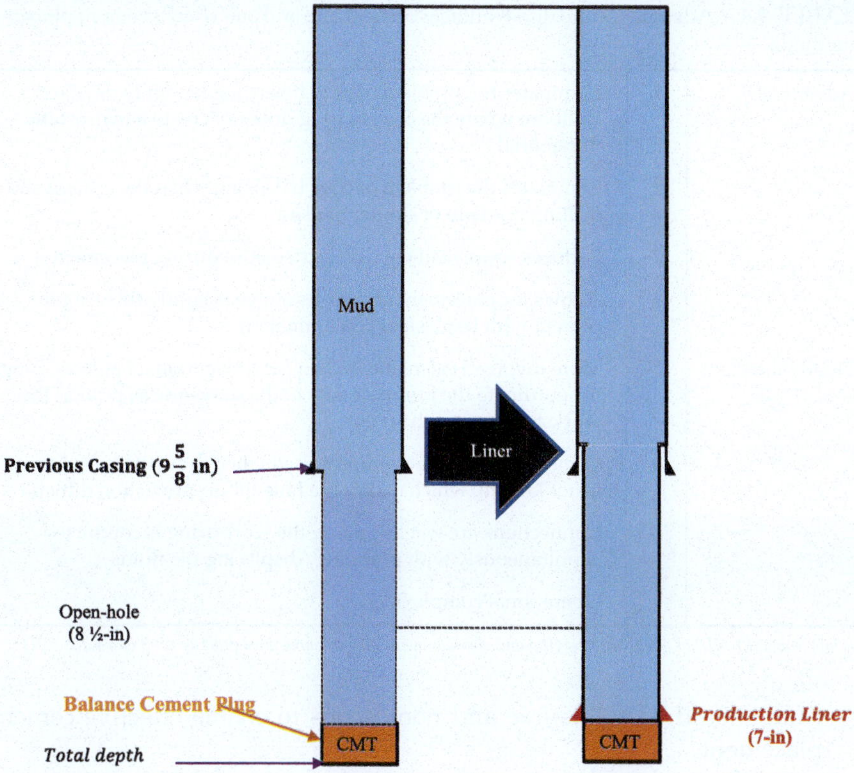

FIGURE 9.3 Balanced cement plug set for off-bottom open-hole completion. First the cement plug is set, and then the liner string is run.

pipe) is used, to prevent U-tubing or free fall of fluids during connections, the annular preventer should be closed.

9.2.3 Through tubing

This method is usually applied only for P&A. When annular cement has been verified above a potential flow zone, it may be possible to place cement through the existing completion string or tubing to create a continuous permanent plug. To achieve that, first the reservoir zone below the packer should have already been sealed by cement squeezing, etc. Next the tubing above the packer is cut, or some holes are created in it. Next, cement slurry is pumped down the completion string back to the annulus (Fig. 9.5).

Some considerations should be made for through-tubing placement such as (API, 2021):

TABLE 9.1 Advantages and disadvantages of the pump and pull (P&P) cement plugging method.

Advantages	1	Eliminates the problem of plug balancing in highly deviated wellbore where the cement plug does not create a hydrostatic differential
	2	Eliminates the problem of plug balancing when the cement and drilling fluid are of similar density
	3	Reduces equivalent circulating densities during placement
	4	Allows for placement of viscous or even slightly thixotropic cement with less risk of contamination
Disadvantages	1	More involved calculations than for conventional balanced plugs to coordinate the pumping rate to the same rate as pulling the workstring out of the hole
	2	Cement plug may be compromised when the open-hole diameter is not known, which makes the hole fill-up prediction difficult
	3	Connections are wet as pulling the workstring is conducted simultaneously with pumping/displacing the fluids
	4	Operational complexity

(Data from API. (2021). Wellbore plugging and abandonment. API recommended practice 65-3 (1st Ed.). API).

- Mud should be solids-free and nonviscous to enable effective cement placement,
- Condition of the existing annular cement,
- Tubing and casing conditions,
- Evidence of sustained casing pressure (SCP), and
- Control lines and gauge cables.

9.2.4 Sacrificial workstring

A sacrificial workstring (tailpipe) should be considered when quick thickening times are needed or when other conditions present a high risk of destabilizing the plug when the workstring is withdrawn. This method is also useful for plugging long intervals, as it eliminates the need for additional time to pull the string above the top of the cement.

Specialized tools are available for tailpipe release, but alternative methods can also be used:

- Shearing off the tailpipe at the end of the job.
- Leaving the string in place until the cement sets, then cutting or backing off the pipe at the first free connection.

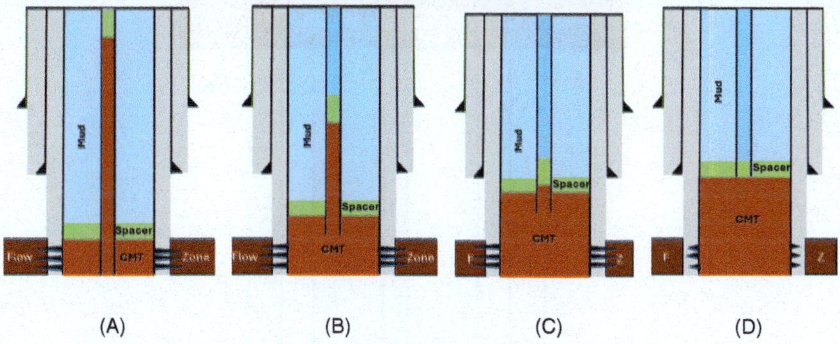

a) **Locate workstring at the base of the cement plug, and a portion of cement is placed into the annulus; typical range can be 30 m (~100 ft) depending on hole size;**
b) **Pull workstring out of the hole at the same rate of the surface pump rate (equivalent annular velocity) between the open-hole and workstring; the cement in the annulus remains at the same height above the base of the workstring;**
c) **Displace until cement is near the balance point of the plug if placed conveniently; the cement in the annulus remains at the same height above the base of the workstring;**
d) **Pull workstring out of the hole and the cement at the top of the plug may become intermixed with the cement spacer or mud.**

FIGURE 9.4 Example of a pump and pull procedure (*Redrawn from API. (2021). Wellbore plugging and abandonment. API recommended practice 65-3 (1st ed.). API*): (A) Locate workstring at the base of the cement plug, and a portion of cement is placed into the annulus; typical range can be 30 m (~100 ft) depending on hole size; (B) pull workstring out of the hole at the same rate of the surface pump rate (equivalent annular velocity) between the open-hole and workstring; the cement in the annulus remains at the same height above the base of the workstring; (C) displace until cement is near the balance point of the plug if placed conveniently; the cement in the annulus remains at the same height above the base of the workstring; and (D) pull workstring out of the hole and the cement at the top of the plug may become intermixed with the cement spacer or mud.

9.2.5 Dump bailer

In this method a small volume of cement slurry is taken via wireline or slickline to the required depth, and then the bailer is opened on impact or by electrical triggering to release the slurry. In this method, usually cement slurry is placed above a mechanical barrier. The advantages of the method are its quick job (because of using line) and accurate control of the depth of plug placement. However, the disadvantages are the limited cement slurry volume that can be transported per run and probable contamination of the slurry during the placement. Despite its limitations, the bailer method is

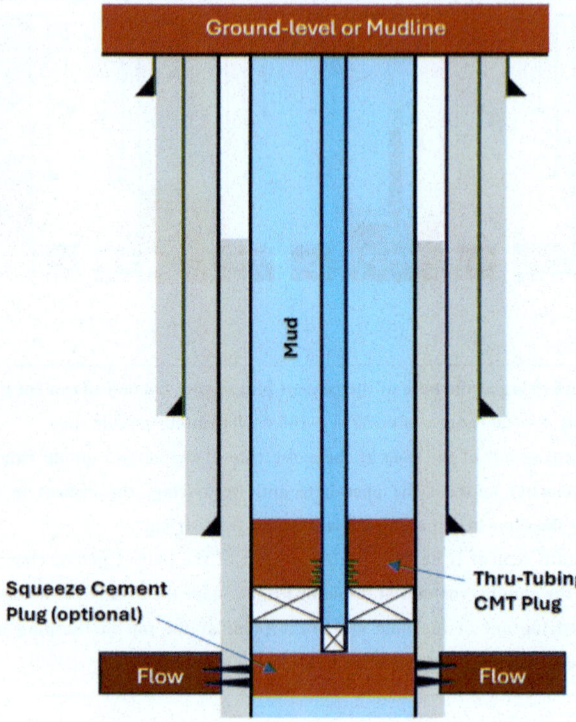

FIGURE 9.5 Through tubing cement squeezing. *(Redrawn from API. (2021). Wellbore plugging and abandonment. API recommended practice 65-3 (1st ed.). API).*

a practical option for setting small, controlled cement plugs in wellbore operations.

9.2.6 Cement squeezing

Cement squeezing is a necessary operation in certain well interventions to improve well integrity and isolate unwanted flow paths. This method is commonly applied in the following scenarios:

- Repair faulty primary cements: If primary cementing could not be done successfully and there is an empty space behind the casing string, secondary cementing is essential, typically done via cement squeezing.
- Shutting-off perforations: When a part of the perforation zones becomes nonhydrocarbon or unusable, there would be water intrusion from some perforations into the well, disturbing the oil and

gas production. To resolve this issue, those zones should be shut off for workover. An efficient method to accomplish this is through cement squeezing, although other methods, like casing patches, are also viable.

- Killing the well (tertiary well control): In very extreme well control conditions, cement slurry may be squeezed or pumped with high pressure into the wellbore against the blowout stream of influx fluids in the well to push influx back to the formation and thereby kill the well.

Fig. 9.6 illustrates the schematic of this operation when nonretrievable packers are used. The process involves the following steps:

1. First a bridge plug is installed at the bottom of perforations to seal the bottomhole from the hydraulic pressures that will be applied.
2. The drill pipe is run in the hole above the perforations.
   a. If a packer is used, it is set above the perforations to isolate the squeeze zone (from the top hole). The packer can be either nonretrievable, like cement retainers, or retrievable (RTTS).
   b. If no packers are used and shallower casings in the hole can tolerate squeezing pressures, close the pipe rams (to seal the top from applied pressures) and then apply squeeze pressures. This is called the *Braden head method*.
3. The injectivity test is essentially conducted by injecting/pressurizing the drilling fluid, to ensure that the formation will later allow cement slurry to be pumped.
4. An adequate volume of cement slurry is spotted against the perforations. Squeezed or pumped with the required pressure.
5. The end of squeezing is indicated by a pressure spike at the surface (drill pipe) which signifies that the perforations are already plugged.
6. At the end of squeezing, postsqueeze operations are made, which depend on the type of packer used:
   a. For nonretrievable packers: Sting-out (special operation done by tools man), which allows the drill pipe to get out of the cement retainer and the valve in the cement retainer to close/seal. At this point the squeezed zone is sealed from the top of the cement retainer. Next, pull the drill pipe a few stands and then make reverse circulation (pumping drilling fluid from the annulus and back from the drill pipes); thus any possible excess cement slurry in the drill pipe can be cleaned to prevent cement setting in the drill pipes (see Fig. 9.6).
   Note: For reverse circulation the surface setup should be ready on the rig floor to let the excess cement return to the disposal pit.

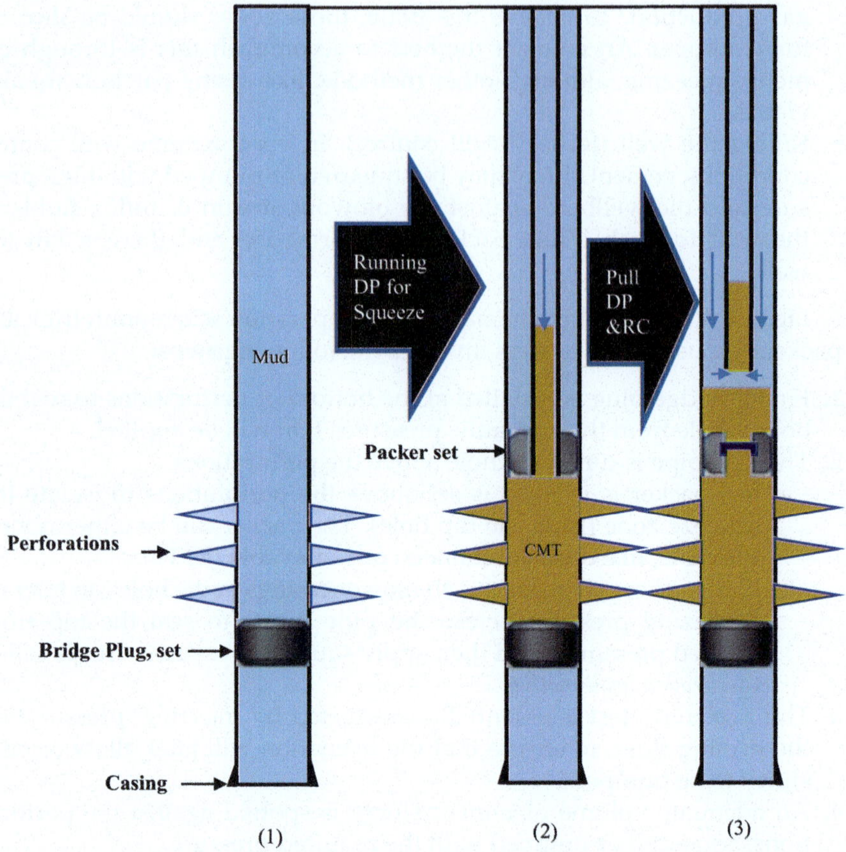

FIGURE 9.6 Stages of cement squeezing for perforation shut-off: (1) setting the bridge plug on the bottom of the perforations, (2) running the drill pipe in the hole, setting packer, spotting large volume cement slurry, and squeezing it into perforations, and (3) stinging-out to allow pulling the drill pipe and the cement retainer's valve closes/seal. At this point, make reverse circulation to clean the drill pipe from possible excess cement and pull the string out of the hole.

b. For retrievable RTTS packers: If no excess cement is in the string, maintain pressure for a waiting-on-cement (WOC) period (typically 15 hours) to allow the cement to set (so that subsequent reverse circulation would not damage the cement plug). After WOC, unset the packer and perform reverse circulation.
Note: Although RTTS packers are cost-effective, as they can be reused, they require additional waiting time, making nonretrievable packers preferable in some cases.

7. Pull the string out of the hole.
8. After waiting on cement for a few hours to let it set, run in the hole with milling or drilling assembly to drill out excess cement and restore wellbore access.

The effectiveness of squeeze cementing depends on the applied pressure: Enough pressure should be applied in squeezing operations for success of operations.

- High-pressure squeeze: Cement is injected above formation fracturing pressures when fracturing is necessary to displace the slurry into the formation.
- Low-pressure squeeze (most common): Cement is pumped below the fracturing pressure, allowing wellbore fluids to be injected into permeable formations such as sand, fractured limestone, or lost-circulation intervals.
- Hesitation squeeze: Used in low-permeability formations, where the cement slurry is pumped intermittently (on and off) to allow gradual infiltration into the formation.

9.2.7 Level-off plugging

The level-off cement plugging method is used to address lost circulation by allowing cement slurry to naturally infiltrate and seal fractures or caverns within a loss zone. This process is carried out as follows:

- Using the drill pipes the RTTS packer is installed inside the casing, ensuring sufficient spacing from the loss zone to minimize operational risks.
- Some cement slurry is spotted above the loss zone in cased hole and with adequate spacing/gap from the packer to prevent complications.
- The slurry is allowed to penetrate the formation under gravity, facilitated by the presence of fractures or caverns that enable natural infiltration and plugging of the loss pathways.
- If the operation is successful the mud level in the drill pipe gradually drops (levels off) until the hydrostatic pressure of the wellbore fluid column equals the formation pore pressure. This controlled pressure balance helps ensure a stable cement plug.
- At the end of the process a predetermined cement height must remain above the loss zone to effectively prevent further fluid loss, ensuring well integrity. Fig. 9.7 shows the initial and final conditions of a typical well using the level-off cement plugging method. For more information, see Ashena et al. (2021).

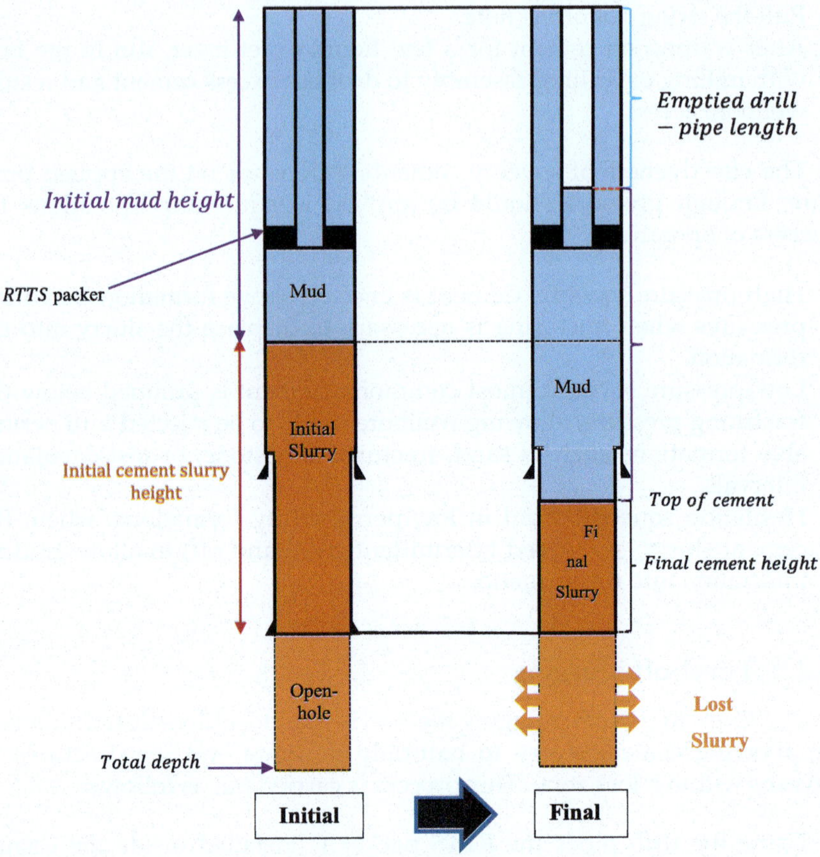

FIGURE 9.7 Level-off cement plugging for putting an end to lost circulation. At the final state (end of the level-off cementing operation), a specified cement height is expected to remain above the loss zone. *RTTS*, Retrievable test-treat-squeeze (Ashena et al., 2021).

Conventional cement is used for plugging nonreservoir thief zones. However, magnesium cement slurries are preferred for plugging reservoir thief zones because they are dissolvable by an HCl-acid wash (in carbonates) or HF wash (in sandstones), to restore permeability and remove any reservoir formation damage.

9.3 Cement plugs evaluation

Evaluating a cement plug after placement is crucial to ensuring its integrity and effectiveness. The following methods are commonly used for assessment after it is set:

9.3.1 Weight test

The weight test is a fundamental method to determine the cement plug's location and assess its initial strength. Compressive strength plays a critical role in the plug's integrity, so sufficient WOC time is required based on laboratory recommendations and wellbore conditions. The procedure is as follows:

- Tag/Dress the plug:
  - Lower the drill string with a mill or bit to tag the cement plug.
  - Compare the actual tag depth with the expected depth. A lower-than-expected tag depth may indicate cement loss or placement issues.
- Initial weight test (low load and circulation):
  - Apply a low weight (2–3 klb) on the cement plug.
  - Maintain a slow circulation rate [e.g., 100 gpm for an 8.5-in hole].
  - Observe returns at the shale shaker:
    If the plug has not fully set, applying additional weight will wash away the top of the plug. Cement fragments in the return flow indicate insufficient setting time.
  - If washing occurs, pull up 1–2 drill pipe stands and continue slow circulation to allow further setting.
- Final strength test (higher load and circulation):
  - After a few hours, rerun the string to tag the plug again.
  - Gradually increase the weight to 10 klb while raising the circulation rate to 300 gpm.
  - Observe pump pressure: a stable pressure indicates good cement integrity, whereas a drop suggests plug failure.
  - Maintain this test for 5–10 minutes to confirm plug stability.

9.3.2 Pressure integrity tests

A hydraulic pressure test verifies the cement plug's ability to withstand downhole pressures.

Procedure:

- Apply 2000 psi of hydraulic pressure on the cement plug.
- Monitor the pressure gauge—any pressure drop indicates potential leakage or plug failure.
- In plug and abandonment (P&A) operations, an inflow test may be performed to confirm plug integrity by checking for fluid movement across the plug.

For more details on hydraulic pressure testing, refer to Chapter 6 "Pressure integrity tests."

References

API. (2010). *API RP 65-2: Isolating potential flow zones in drilling and well completion operations* (2nd ed.). API.

API. (2021). *Wellbore plugging and abandonment. API recommended practice 65-3* (1st ed.). API.

Ashena, R., Ghalambor, A., Elmgerbi, A., Hekmatinia, A.-A., & Mubashir, M. (2021). 'Level-off' cement plugging method to cure lost circulation verified with case studies. *Journal of Petroleum Exploration and Production Technology, 11*, 2777–2789. https://doi.org/10.1007/s13202-021-01177-3.

Ashena, R., Bahreini, H., & Lotfi, M. (2023). Statistical analysis of past kicks and blowouts occurred in a Middle Eastern oilfield. *Journal of Petroleum Exploration and Production Technology, 13*, 2063–2082. https://doi.org/10.1007/s13202-023-01664-9.

Ashena, R. (2025). *The practical handbook of well control*. Routledge, Taylor and Francis Group https://www.routledge.com/The-Practical-Handbook-of-Well-Control/Ashena/p/book/9781032753850?srsltid=AfmBOooZqOJyBoNQ8lWhWyMcNMmdX4-St1TUa2goxJzlN_6-31cO4lBo.

Nelson, E., & Guillot, D. (2006). *Well cementing* (2nd ed.). Schlumberger.

CHAPTER

10

Data-driven and machine learning approaches to cement evaluation

Jocin J. Abraham[a], *Mahmood Amani*[a], *and Catalin Teodoriu*[b]

[a]**Texas A&M University at Qatar, Doha, Qatar,** [b]**University of Oklahoma, Norman, OK, United States**

10.1 Introduction

Understanding the behavior of oilwell cement can be challenging owing to the wide array of properties that need to be considered and parameters involved in evaluating each of these. Factors, such as cement formulations and additives used, mixing conditions, curing time, and curing conditions, play a key role in determining the development of mechanical properties of the cement. All these parameters and variables are readily measured and available both from the field and from laboratory-based experimentation. Elmgerbi et al. (2022) presented a data-driven approach to evaluating and presenting solutions for drilling problems, but no machine learning (ML) methods were applied. However, although a large amount of data is already available, much analysis has not been conducted to derive meaningful insights into cement behavior beyond the most common properties of short-term uniaxial compressive strength (UCS) and cement bonding.

ML algorithms can be utilized to develop data-driven models, which can help in discovering useful patterns and trends from the available data, as well as help in predicting properties, which may not usually be easily obtained. ML models can also help in removing human bias in cement log interpretations while providing a faster, efficient, and reliable method to estimate cement properties without resorting to the time-consuming experimental or destructive processes. Different ML algorithms can be

Well Cement Integrity
DOI: https://doi.org/10.1016/B978-0-443-51564-4.00010-2

applied for different tasks, and their effectiveness depends on the dataset and parameters used for training the models. A common ML model is the artificial neural network (ANN), applicable to petroleum engineering applications (Ashena & Thonhauser, 2015). Depending on the application, algorithms applicable for cementing applications include those for clustering and classification, image, and pattern recognition, as well as algorithms for predicting cement properties.

The primary objective of this chapter is to explore the possibilities and potential of utilizing ML algorithms to aid in cement evaluation. Discussed below are a few case studies that highlight the application of ML algorithms and other data-driven approaches to better understand the properties and behavior of oilwell cement and simplify the process of selecting the optimal cement formulation for any oil and gas well.

10.2 Cement bond evaluation

Generally, cement evaluations are performed from measurements of acoustic parameters, such as acoustic impedance (AI), acoustic wave attenuation, refracted wave analysis, wave arrival times, wave amplitude measurements, or a combination of these, and are quite subjective, as interpretations are usually performed by engineers or petrophysicists. Different ML techniques can be utilized to reduce human bias and better interpret various cement bond logs (CBLs) faster, more accurately, and in a more cost-effective manner. ML algorithms can account for and evaluate acoustic data quality, casing–formation–cement bonding, cement spread, and zonal isolation and identify channeling and the presence of microannuli (Voleti et al., 2020).

Voleti et al. (2020) utilized several supervised ML approaches, such as random forest (RF), logistic regression (LR), *k*-nearest neighbor (KNN), and support vector machine (SVM), to evaluate cementing quality. The output and inputs used in the training and testing of ML models are cement evaluation flags—cement quality flag calculated using bond index (BI), which measures wave attenuation, and variable density log (VDL) flags calculated using waveform arrival times. BI was classified as poor bond (BI < 60%), moderate bond (60% > BI < 80%), and good bond (BI > 80%), whereas VDL was classified as free pipe, fast formation, and non-free pipe. The trained models were also used to perform blind tests on five wells, and the results obtained showed high accuracy and were in good agreement with evaluations made by humans.

Viggen et al. (2020) proposed interpreting cement evaluation logs using supervised deep convolutional neural networks (CNNs) to classify bond quality (BQ) and hydraulic isolation (HI)—a task that was previously interpreted manually by trained professionals. CNNs are very capable when

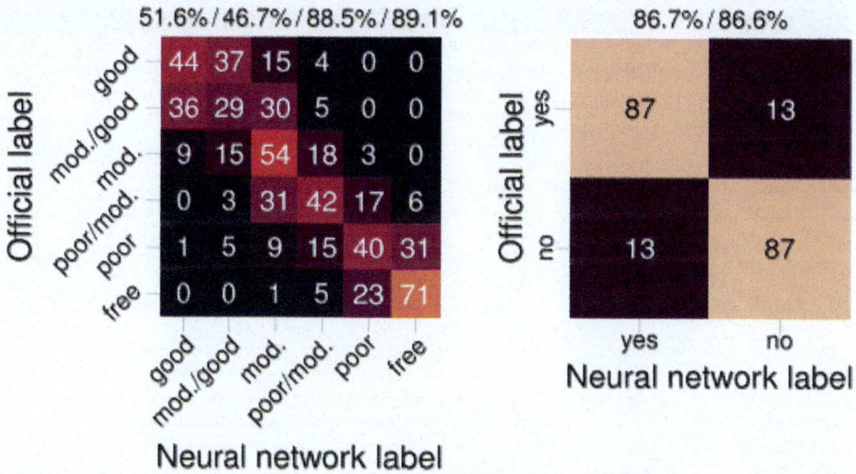

FIGURE 10.1 Confusion matrices for the neural network for bond quality (*left*) and hydraulic isolation (*right*), used to train the models (Viggen et al., 2020).

it comes to image classification, and the task of interpreting well status using well log analysis from a collection of images and log curves falls broadly under general image classification. The input curves used for the neural network included corrected cement bond log (CBLF), sonic imaging using VDL, six Ultrasonic Imaging Tool channels (ECCE/AZEC, AWBK, IRAV, IRBK, T2BK, AIBK, UFLG, and UCAZ/RB), and the gamma ray log. The outputs were BQ (ranging from good, moderate to good, moderate, moderate to poor, poor, and free pipe) and HI (yes or no/uncertain). To train the model, confusion matrices were drawn up for the different outputs of BQ and HI based on official interpretation and the predictions by the neural network, as shown in Fig. 10.1. Four accuracy metrics were used to ensure that the model predictions were reliable, and overall results indicated rather promising results with over 50% accuracy (Viggen et al., 2020). Further work done jointly by Viggen et al. (2023) and Viggen (2022) built on this and looked closer at assisted cement log evaluation using ML models and created an open-source tool for interpreting cement bond using high-quality training datasets. This model can further classify annular condition of the cement layer, as shown in Fig. 10.2.

CNNs and pattern recognition were also used by Imrie (2021) to determine cement channeling and the cement BQ in oil and gas wells. An azimuthal bond map from an ultrasonic cement evaluation log and a CBL density waveform are used as the inputs, and the ML model can classify the BQ into narrow channeling, wide channel, very wide channel, and partial bonding using pattern recognition. Over 200,000 ft of bond log data gathered from real logs and data augmentation was used for training the

Code	Isolating potential	Description
CEM 1A	High	Well bonded homogeneous cement around the entire annulus
CEM 1B	High	Well bonded heterogeneous cement around the entire annulus
CEM 1C	High	Well bonded heterogeneous cement around the entire annulus with non-connected pockets or short liquid filled channels
CHNCem 2A	Low	Continuous liquid filled channeling of <20% width
CHNCem 2B	Low	Continuous liquid filled channeling of 20–40% width
CHNCEM 2C	Low	Continuous liquid filled channeling of >40% width
CONCem 3A	High	Well bonded cement on high side, slightly contaminated cement on low side
CONCem 3B	Medium	Well bonded cement on high side, heavily contaminated cement on low side
CONCem 3C	Medium	Homogeneous contaminated (or unset) cement with lower impedance than expected
PATCem 4A	Medium	Patchy cement bond providing medium isolation potential
PATCem 4B	Low	Patchy cement bond providing low isolation potential
MAWCem 4C	High	Wet microannulus providing high isolating potential
MAWCem 4D	Medium	Wet microannulus providing medium isolating potential
MAWCem 4E	Low	Wet microannulus providing low isolating potential
MADCem 4F	Low	Dry microannulus
GCCem 5	Low	Gas cut cement
LCCem 6	Low	Cement in liner lap with eccentralised liner
FORM 7A	High	Formation bond of good quality (barrier-level)
FORM 7B	Medium	Formation bond of medium quality (not barrier-level)
FORM 7C	Low	Formation bond of low quality (nor barrier-level)
FORMCEM7C	Low	Combined cement and formation bond with poor bond
FORMCEM7D	Medium	Combined cement and formation bond with medium bond
FORMCEM7E	High	Combined cement and formation bond with good bond
MUDS 8A	None	High-density settled mud solids, good bond
MUDS 8B	None	Medium-density settled mud solids, medium bond
MUDS 8C	None	Low-density settled mud solids, patchy bond
FPL 9A	None	Free pipe filled with liquid
FPG 9B	None	Free pipe filled with gas or light oil
OTHER	—	An annulus status that does not fit the above
PDQ	—	Poor data quality, rendering the data uninterpretable

FIGURE 10.2 Annular condition flags following machine learning evaluation of cement logs using convoluted neural network model (Viggen, 2022).

model, whereas over 45,000 ft of log data was used for blind validation testing. Although these pattern-matching models yield high accuracy and can reduce bias in interpretation, a major limitation is the fact that accuracy depends on if the acquisition data finds a similar log image match in the library (Imrie 2021).

Itikawa et al. (2023) evaluated several different ML algorithms and their efficiency in aiding in automatic interpretation of cement logs. ML models were created using classification algorithms, including KNN, Decision Tree, RF, gradient tree boosting, SVM, LR, and Gaussian Naïve Bayes as well as neural networks. A database of cement logs from 25 wells, which corresponded to around 20 km of interpreted logs, was utilized to train and test the different ML models. These logs included acoustic logs, such as CBL, VDL, and ultrasonic logs. Feature sets for each of the ML models were created using a combination of the logs, and the log response was classified into BQ classes ranging from free pipe (Class 1), poor bond (Class 2), possible isolation (Class 3), probable isolation (Class 4), and good bond (Class 5). Evaluation of the different models indicated that the RF model has the best overall performance and better accuracy in evaluating cement bond from log data (Itikawa et al., 2023).

10.3 Cement integrity evaluation in plugging and abandonment

At the end of the lifecycle of a well the plugging and abandonment (P&A) operation is performed to ensure well integrity. To verify that the wellbore is safely and permanently isolated from the surrounding formation and geology, logging operations are conducted. However, vast amounts of logging data are generated from these jobs, which require complex and time-consuming analysis with significant chances of error in the assessment. ML techniques can significantly reduce the time required in the analysis of cement bond logging data generated from tools such as CBL, VDL, and BI while improving the reliability and accuracy of the results.

Camerini et al. (2023) evaluated three different unsupervised ML clustering algorithms—*K*-means, bisecting *K*-means, and Gaussian mixture model—to evaluate logging data obtained from P&A operations using a P&A assistant software. Input parameters for the feature sets were obtained from the CBL channel, BI channel, and VDL channel as well as the AI (AIBK) channel. Performance and reliability of these unsupervised ML algorithms were evaluated using intrinsic clustering metrics of silhouette coefficient, which has a scoring range of [−1,1], and rand index, which has a scoring range of [0,1]. From their evaluations, it was noted that *K*-means algorithm generated the most reliable model and was closely followed by the Gaussian Mixture model in overall performance. Using these clustering algorithms and ML models can aid in faster interpretation and classification of BQ and cement integrity while reducing human error and interpretation bias (Camerini et al., 2023).

Bose et al. (2021) also utilized deep learning algorithms to aid in the evaluation of cement quality behind multiple casing strings—especially during P&A operations—where regulations require primary and secondary (cement) barriers to be established across the entire cross-section of the wellbore. Current ultrasonic and other acoustic tools face limitations in their depth of investigation, especially in the presence of multiple (inner and outer) casing strings. Ultrasonic measurements, which probe shallow near wellbore (inner annulus), and lower frequency sonic measurements, which probe deeper into formation (outer annulus and formation), were utilized to generate the datasets required for the ML models. This acoustic response data is used as input and is utilized to collectively predict AI and flexural attenuation of the outer casing using deep learning neural network models (Bose et al., 2021).

10.4 Prediction of compressive strength

Given the importance of cement in the well integrity matrix, understanding its mechanical properties and their behavior over time is of great importance. Among these mechanical properties, compressive strength (CS) of cement is of utmost importance because it directly relates to the stability of the wellbore, operational lifecycle of the well, and the integrity assessment during operations. Though CSs of cement are usually obtained through physical experiments in the lab or field, ML approaches can be utilized to determine cement CSs to great accuracy. Although there have not been many studies on the application of ML to predict CSs for oilwell cements, a lot of work has been done for concrete in the construction industry (Chou, et al., 2014; Feng et al., 2020; Paudel et al., 2023; Shrivastava & Shrivastava, 2024; Yan & Shi, 2010), which has demonstrated the use of different ML algorithms to predict CS. Similar approaches can be utilized for oilwell cement to predict their CSs using ML approaches rather than costly and time-consuming physical experiments. Work done by Nyakilla et al. (2022, 2023) and Abraham et al. (2023) has demonstrated different approaches that could be used for predicting mechanical properties such as CS of oilwell cement.

Nyakilla et al. (2022) used experimental data from oilwell cements and ML algorithms to build a model that could predict CS and shear bond strength (SBS) of cement over time. Four main algorithms were tested—Gaussian process regression (GPR), gradient boost regression trees (GBRTs), SVM, and ANNs. Data for the training and testing datasets were obtained from experimental evaluation. Oilwell cement composition, including water content, base cement weight, fly ash content, dispersants, and curing time was used as data input for the ML algorithms, whereas SBS and CS were set as outputs. As shown in Fig. 10.3, models generated using GPR, SVM, and ANN showed reasonable accuracy in predicting both CS and SBS, whereas the highest accuracy was noted for the GBRT model. Based on input data the GBRT model can accurately and successfully estimate the CS and shear bond of oilwell cement. For both the CS and SBS strengths the model showed the least mean square error, indicating that the model's accuracy has remained high and that the prediction error is quite small (Nyakilla et al., 2022).

Abraham et al. (2023) utilized experimental data and ML algorithms to evaluate UCSs of different formulations of oilwell cements. ML models

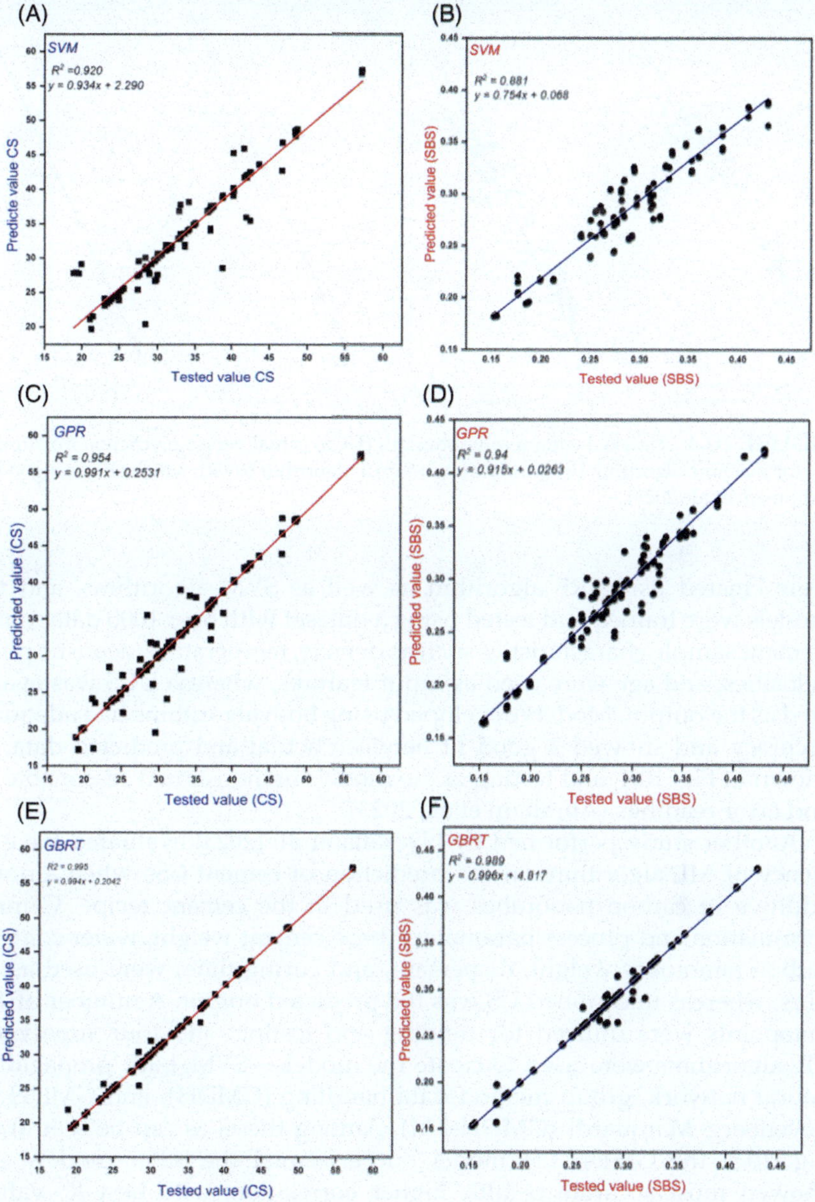

FIGURE 10.3 Experimental and predicted values of shear bond strength (SBS) and compressive strength from different algorithms. (A, B) SVM, Support Vector Machines (C, D) GPR, Gaussian Process Recession (E, F) GBRT, Gradient boost regression tree (Nyakilla et al., 2022).

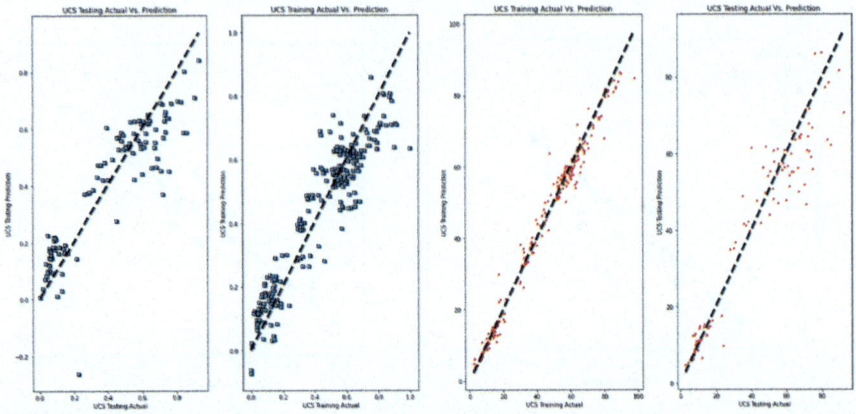

FIGURE 10.4 Uniaxial compressive strength (UCS) actual versus prediction for support vector machine algorithm (*left*) and random forest algorithm (*right*), which shows a good fit (Abraham et al., 2023).

were created using RF algorithm as well as SVM algorithms, and the models were trained and tested using a dataset with over 1000 datapoints. Cement sample characteristics such as density, temperature, acoustic pulse velocities, and age were used as input features, whereas UCS was evaluated as the output. Models developed using both algorithms had adequate accuracy and showed a good fit between actual and predicted data, as shown in Fig. 10.4, and testing of the models demonstrated reasonable R^2 and error readings (Abraham et al., 2023).

Another study performed by Nyakilla et al. (2023) evaluated the efficiency of ML algorithms in the prediction of cement CSs when a novel additive of carbon nanotubes was used in the cement recipe. Cement formulation and process parameters (base cement weight, water content, carbon nanotubes weight, dispersant, and curing time) were used as inputs, whereas unconfined CS was the predicted output. A number of 150 datapoints were utilized for training and testing, and four supervised ML algorithms were used to create the models—SVM, back propagation neural network, group method data handling (GMDH), and GMDH by Levenberg–Marquardt (GMDH-LM). Among these, as can be seen from Fig. 10.5, the GMDH-LH model outperformed the other models and showed most accurate results, higher correlation, with best R^2 values and lowest errors. Curing time had the largest influence on the cement mechanical strength, followed by the cement composition (including the nanotubes). From this, it could be noted that algorithms, such as GMDH-LH, are more robust and can be utilized to reliably predict cement CSs when the composition of oilwell cement is tweaked without resorting to experimental analysis.

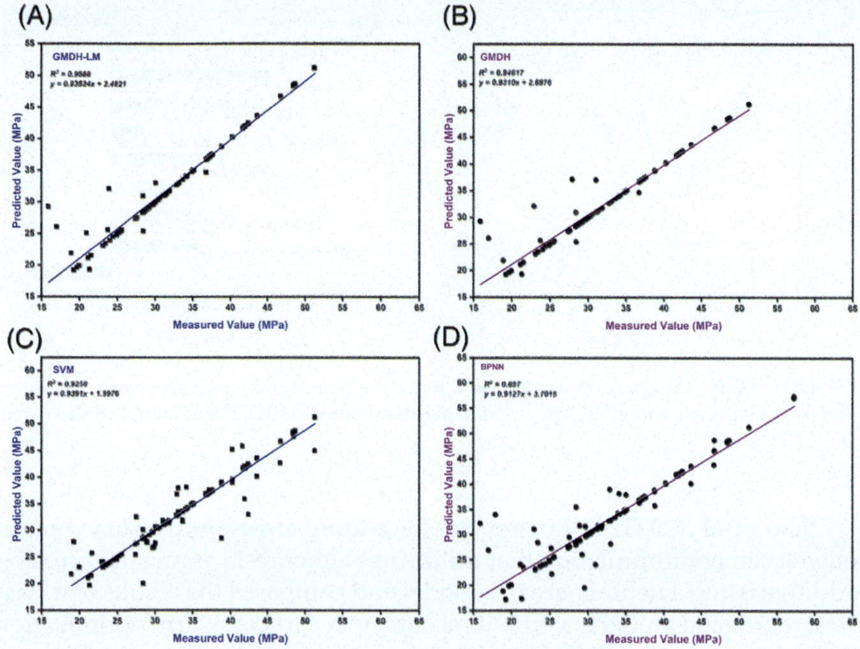

FIGURE 10.5 Measured and predicted compressive strength for different algorithms when tested for carbon nanotube-infused cements. (A) GMDH-LM, group method data handling by Levenberg Marquardt; (B) GMDH, group method data handling; (C) SVM, support vector machine; (D) BPNN, Back propagation neural network (Nyakilla et al., 2023).

10.5 Corrosion evaluation of cements using neural networks

Oil and gas wells are very often drilled in highly corrosive and challenging formations with high pressure and high temperature conditions, which can be prone to well integrity issues (Abid et al., 2023). Oilwell cements utilized in such wells need to be adapted to meet the challenges of such hostile downhole environments and should be resistant to damage and corrosion caused by the water salinity or by the presence of organic acids or acidic gases such as carbon dioxide (CO_2) or hydrogen sulfide (H_2S). As cement experiences a high degree of corrosion in acidic environments, additives such as corrosion inhibitors, pH agents, or chemical scavengers such as sodium sulfite and sulfur dioxide are commonly utilized to improve the corrosion resistance of oilwell cement. However, as there are a lot of cement formulations, the degree of long-term corrosion is often difficult to evaluate, which can be remedied using data-driven approaches.

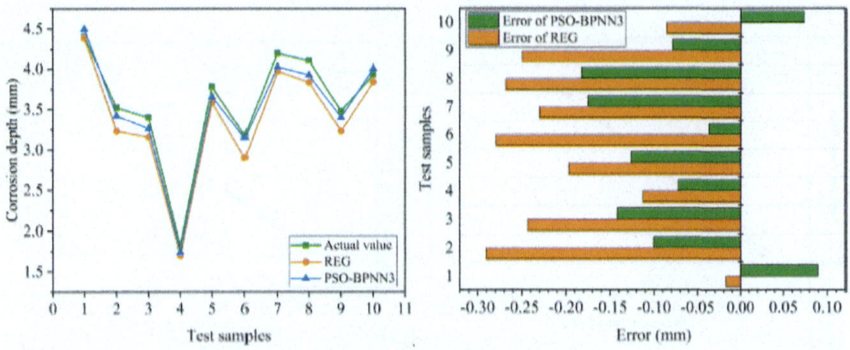

FIGURE 10.6 Comparison of corrosion depth predicted by the particle swarm optimization by back propagation (PSO-BP) model, regression model (REG), and actual values (Zhao et al., 2023).

Zhao et al. (2023) evaluated the long-term corrosion resistance of an oilwell cement formulation that utilized polymer resin as an anticorrosion additive using a neural network model and compared the results to a classical regression model to verify its accuracy. A particle swarm optimization by back propagation (PSO-BP) algorithm was utilized to generate the main corrosion prediction model based on experimental data obtained from laboratory testing of anticorrosion cement slurry with polymer resin additive in CO_2 environments. The model relied on three main input parameters—temperature, CO_2 pressure, and time—whereas the output parameter was the depth of corrosion of the cement (reduction in thickness). Over 100 sample groups were obtained from laboratory experiments, which were utilized to train, test, and validate the model.

Four PSO-BP models were generated by Zhao et al. (2023) by varying the number of hidden layers in the neural network, and all the prediction models had high accuracy between the training and testing datasets. However, three of the models had significant validation and testing errors, indicating the importance of optimization of the PSO-BP algorithm in model creation. Results obtained from the PSO-BP models were used to predict the long-term corrosion of the oilwell cement with the polymer resin additive, and the results were compared to the predictions from a classical regression algorithm. Based on the model testing and errors (Fig. 10.6), PSO-BP model had higher accuracy, lower deviation, and smaller error in the prediction of the depth of corrosion in the long term compared to the regression models (Zhao et al., 2023). From these, applications of such ML models can help in predicting and evaluating long-term corrosion degree of oilwell cements and can help guide the design of anticorrosion cement formulations.

10.6 Summary

The use of AI and ML-driven methods in cement bond and quality evaluation is rapidly growing as the industry faces challenges such as diminishing experience and the need for cost-effective well management. These technologies offer the potential for continuous, real-time cement evaluation, providing valuable insights that support more efficient and data-driven decision-making. By analyzing large datasets, AI and ML can identify patterns and factors influencing cement performance that may not be immediately evident through traditional methods.

However, the effectiveness of AI and ML in cement evaluation is heavily reliant on the availability of high-quality, meaningful data. Although these technologies have the potential to significantly enhance cement assessment, ongoing challenges include identifying the key factors that influence cement integrity and ensuring data quality.

References

Abid, K., Srivastava, S., Romero, M., Amani, M., & Teodoriu, C. (2023). Experimental and machine learning study of thermal conductivity of cement composites for geothermal wells. *Geothermics, 110*, 102659. https://doi.org/10.1016/j.geothermics.2023.102659.

Abraham, J. J., Devers, C., Amani, M., & Teodoriu, C. (2023). Data driven approach to evaluating cement sheath properties in existing and abandoned wells for field redevelopment. In *Gas & Oil Technology Showcase and Conference*. Society for Petroleum Engineers https://doi.org/10.2118/214222-MS.

Ashena, R., & Thonhauser, G. (2015). Application of artificial neural networks in geoscience and petroleum industry. In C. Cranganu, H. Luchian, & M. Breaban (Eds.), *Artificial intelligent approaches in petroleum geosciences*. Springer https://doi.org/10.1007/978-3-319-16531-8_4.

Bose, S., Zhu, L., Zeroug, S., Merciu, I. A., Constable, K., Hemmingsen, P. V., Berg, E., Erik, W., Govil, A., & Kalyanaraman, R. S. (2021). Acoustic evaluation of annulus B barriers through tubing for plug and abandonment job planning. In *Offshore Technology Conference*. Society of Petroleum Engineers https://doi.org/10.4043/31302-MS OTC-31302-MS.

Camerini, I. G., Ferreira, G. R. B., de Souza, L. P. B., Hidalgo, J. A. S., Correia, T. M., Rodrigues, A. S., & Batista, J. H. G. (2023). Machine learning assisted cement integrity evaluation during pluggingand abandonment operations. In *Abu Dhabi International Petroleum Exhibition and Conference*. Society of Petroleum Engineers https://doi.org/10.2118/216950-MS.

Chou, J. S., Tsai, C. F., Pham, A. D., & Lu, Y. H. (2014). Machine learning in concrete strength simulations: Multi-nation data analytics. *Construction and Building Materials, 73*, 771–780.

Feng, D. C., Liu, Z. T., Wang, X. D., Chen, Y., Chang, J. Q., Wei, D. F., & Jiang, Z. M. (2020). Machine learning-based compressive strength prediction for concrete: An adaptive boosting approach. *Construction and Building Materials, 230*, 117000. https://doi.org/10.1016/j.conbuildmat.2019.117000.

Elmgerbi, A., Les, B., Ashena, R., & Atkin, T. (2022). A practical decision tool to evaluate and rank potential solutions for expected downhole drilling problems during the well-planning phase. *Journal of the Institution of Engineers (India): Series D, 103*, 25–36. https://doi.org/10.1007/s40033-021-00325-7.

Imrie, A. (2021). The application of pattern recognition and machine learning to determine cement channeling and bond quality from azimuthal cement bond logs. In *SPWLA 62nd Annual Logging Symposium. SPWLA-2021-0068*. Society of Petrophysicists and Well Log Analysts 10.30632/SPWLA-2021-0068.

Itikawa, M. A., Ahón, V. R. R., Souza, T. A., Carrasco, A. M. V., Neto, J. C. Q., Gomes, J. L. S., Cavalcante, R. R. H., Ribeiro, I. B., Rocha, J. M. S., Carvalho, C. P. C., & Lima, L. B. (2023). Automatic cement evaluation using machine learning. In *Offshore Technology Conference Brasil. OTC-32961-MS*. Society of Petroleum Engineers https://doi.org/10.4043/32961-MS.

Nyakilla, E. E., Gu, J., Charles, G., Ricky, E. X., Hussain, W., Iqbal, S. M., Kalibwami, D. C., Alareqi, A. G., Shaame, M., & Ngata, M. R. (2023). Application of novel machine learning in the prediction of compressive strength of oil well cement with reinforced nanomaterial. *SSRN Electronic Journal*. https://doi.org/10.2139/ssrn.4296915.

Nyakilla, E. E., Gu, J., Kasimu, N. A., Robert, E. F., Innocent, N., Mohamedy, T., Shaame, M., Ngata, M. R., & Mabeyo, P. E. (2022). Application of machine learning in the prediction of compressive, and shear bond strengths from the experimental data in oil well cement at 80°C. Ensemble trees boosting approach. *Construction and Building Materials, 317*, 125778. https://doi.org/10.1016/j.conbuildmat.2021.125778.

Paudel, S., Pudasaini, A., Shrestha, R. K., & Kharel, E. (2023). Compressive strength of concrete material using machine learning techniques. *Cleaner Engineering and Technology, 15*, 100661. https://doi.org/10.1016/j.clet.2023.100661.

Shrivastava, S., & Shrivastava, T. (2024). Prediction of concrete's compressive strength using machine learning algorithms. *Materials Today: Proceedings, 103*, 183–189. https://doi.org/10.1016/j.matpr.2023.08.252.

Viggen, E. M., Singstad, B. J., Time, E., Mishra, S., & Berg, E. (2023). Assisted cement log interpretation using machine learning. *SPE Drilling and Completion, 38*(02), 220–234. https://doi.org/10.2118/209529-PA.

Viggen, E.M., 2022. New adventures in assisted well log interpretation. https://erlend-viggen.no/new-adventures-in-assisted-well-log-interpretation/

Viggen, E. M., Merciu, I. A., Løvstakken, L., & Måsøy, S. E. (2020). Automatic interpretation of cement evaluation logs from cased boreholes using supervised deep neural networks. *Journal of Petroleum Science and Engineering, 195*, 107539. https://doi.org/10.1016/j.petrol.2020.107539.

Voleti, D. K., Reddicharla, N., Guntupalli, S., Reddy, R., Vanam, P. R., Khanji, M. S., Sumaidaa, S. A., & Ali, M. A. S. (2020). Smart way for consistent cement bond evaluation and reducing human bias using machine learning. In *Abu Dhabi International Petroleum Exhibition & Conference*. Society of Petroleum Engineers https://doi.org/10.2118/202742-MS.

Yan, K., & Shi, C. (2010). Prediction of elastic modulus of normal and high strength concrete by support vector machine. *Construction and Building Materials, 24*(8), 1479–1485.

Zhao, J., Chen, R., Liu, S., Zhou, S., Xu, M., & Dai, F. (2023). Corrosion degree evaluation of polymer anti-corrosive oil well cement under an acidic geological environment using an artificial neural network. *Polymers, 15*(22), 4441. https://doi.org/10.3390/polym15224441.

CHAPTER

11

Plug and abandonment

Rahman Ashena, William K. (Bill) Ott, and William Bellenger

Smart Energy Data Analytics (SEDA)-Group, Houston, TX, United States

11.1 Introduction

Well abandonment has been a longstanding aspect of well construction throughout the history of hydrocarbon exploration. What is changing, however, is the increasing number of wells and fields that are currently inactive, suspended, or nearing the end of their productive life. As abandonment was not considered a crucial part of well life, there have not been major advances in abandonment methods, materials to be used and even policies. However, changes are currently underway. Globally, governments and regulatory bodies are now urging the oil and gas industry to properly seal and permanently retire unproductive wells to prevent environmental impact, by using better and more innovative sealing methods and materials.

In some mature basins the decommissioning of aging infrastructure has resulted in a substantial increase in the number of wells requiring abandonment. These projects are now being treated as independent endeavors rather than the responsibility of existing asset teams and contract structures. This significant potential business opportunity for service companies could lead to the development of specialized tools aimed at minimizing economic impacts for operators while ensuring compliance with abandonment obligations.

11.2 Overview and consequences of lack of plug and abandonment integrity

Well abandonment has been a longstanding aspect of well construction throughout the history of hydrocarbon exploration. What is changing,

Well Cement Integrity
DOI: https://doi.org/10.1016/B978-0-443-51564-4.00014-X

however, is the increasing number of wells and fields that are currently inactive, suspended, or nearing the end of their productive life. Globally, governments and regulatory bodies are urging the oil and gas industry to properly seal and permanently retire unproductive wells to prevent environmental impact.

In some mature basins the decommissioning of aging infrastructure has resulted in a substantial increase in the number of wells requiring abandonment. These projects are now being treated as independent endeavors rather than the responsibility of existing asset teams and contract structures. This significant potential business opportunity for service companies could lead to the development of specialized tools aimed at minimizing economic impacts for operators while ensuring compliance with abandonment obligations.

There are about 30 million abandoned wells worldwide with emissions of 2.5 million tons of methane per year—the climate-damage equivalent of three weeks of US oil consumption (Groom, 2020). Apparently, at least 10% of these wells are in the United States. These wells may leak greenhouse gases and dangerous chemicals into water supplies, air, or soil and, in some extreme cases, cause explosions. Millions of people in the United States live near abandoned oil and gas wells linked to explosions and toxins (NBC, 2023). Therefore such wells pose environmental and health risks to the community.

Orphan wells are those wells whose owners are not known (unidentified owners). In most cases, these wells have not usually been abandoned properly, and thereby they pose a higher risk of environmental incidents. Fig. 11.1 shows a map of more than 120,000 documented orphan wells that are eligible for closure funding by the US federal government. The United States government committed $4.7 billion through infrastructure and Jobs Act of 2021 (Environmental Defense Fund (EDF), 2023). There is some uncertainty about the actual number of undocumented orphan wells, one reason for which is not well knowing locations of such wells. Possibly the actual number of undocumented orphan wells may exceed 1 million (Washington Post, 2024).

Abandonment of oil and gas wells, which can no longer produce, has been an integral goal for the oil and gas industry (Liversidge et al., 2006). It consists of setting several cement plugs with the purpose of mitigating the hazards associated with unplugged abandoned wells, such as groundwater contamination, groundwater comingling, loss of aquifer pressure head, and uncontrolled gas migration.

A plug and abandonment (P&A) job is expected to fulfill the following objectives:

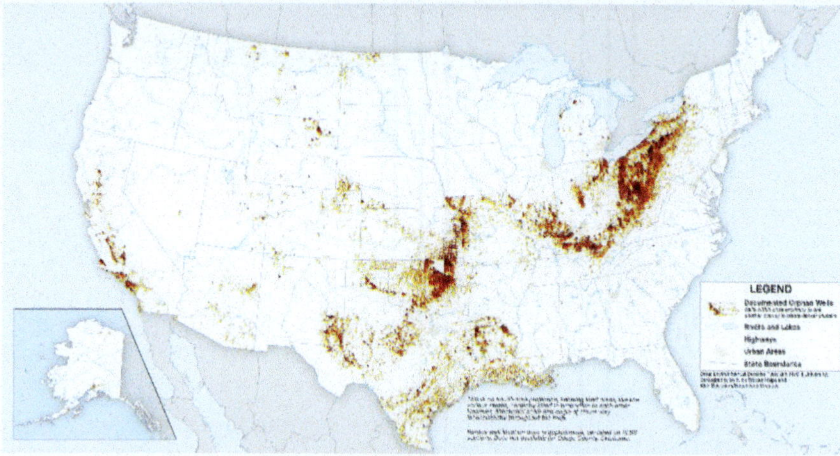

FIGURE 11.1 Map of more than 120,000 documented orphan wells eligible waiting for closure funding. *(Redrawn from Environmental Defense Fund (EDF). (2023). Plugging orphan wells across the United States. Retrieved November 9, 2023, from https://www.edf.org/orphanwellmap).*

- Prevent unacceptable flow of reservoir fluids or hydrocarbons to the surface environment or to another low-pressure formation called underground blowout.
- Eliminating or minimizing waste and CO_2 emissions from the reservoir formation to the environment.
- Protecting the health of the public and avoiding any interference with future human activities so that human life above abandoned wells is not disturbed.
- Remove liabilities from the financial balance sheet, and with a long-term perspective.
- Complying with P&A regulations depending on each country or region's regulatory agency.

The cost of P&A ranges from 76k USD to 1 million USD; P&A costs are 9% greater for gas wells and increase with depth. In another case study the P&A cost for a 5000-m (16,400 ft) well in Germany rose to 1.27 million USD (Osundare et al., 2018). For the whole decommissioning of offshore platforms, well abandonment cost accounts for 40%–45% of all the decommissioning cost in most O&G companies. Governments, like Norway, which takes 78% of oil revenues in tax, must also bear an equally large share of the costs involved in abandonment and decommissioning (Dragland, 2024).

For a two-dimensional view, Fig. 11.2 shows the consequence of an improperly executed P&A job showing underground hydraulic

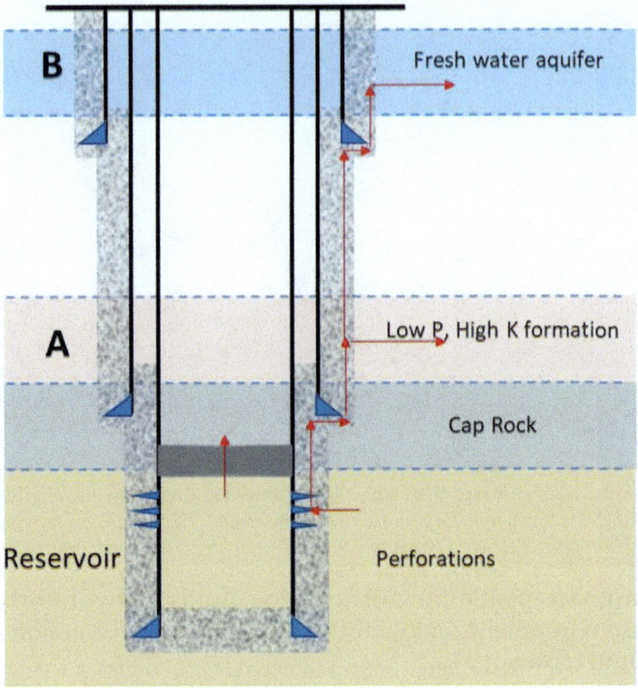

FIGURE 11.2 Schematic of a well with badly executed P&A causing no cement integrity and hydraulic communication between the reservoir and other zones.

communication from the reservoir to possible zones of B (low-pressure formation) and zone A (freshwater zones). In a three-dimensional view, Fig. 11.3 shows possible hydraulic pathways in an abandoned well from the reservoir to the surface because of failure of the annular cement sheath or the cement plug. These two figures illustrate the importance of proper execution and evaluation of cement integrity in P&A wells.

Bad well abandonment can have very serious consequences. As an example, Fig. 11.4 shows a newspaper report of fires related to the blowout of poorly abandoned wells in the Yaggy field, in downtown Hutchinson, Arkansas, destroying two businesses, including a Decor Wedding and Party Supplies store. Several old, abandoned wells blew out in this incident, linked to gas storage operations up to 7 mi (~11 km) away. There were two fatalities in this incident (Mr. and Mrs. Hahn). Because of continuous gas feed from underground the fire extinguishers reported that the fire could not be put out. First, it was assumed that the fire had occurred because of a pipeline leak. However, the source of the incident was bad

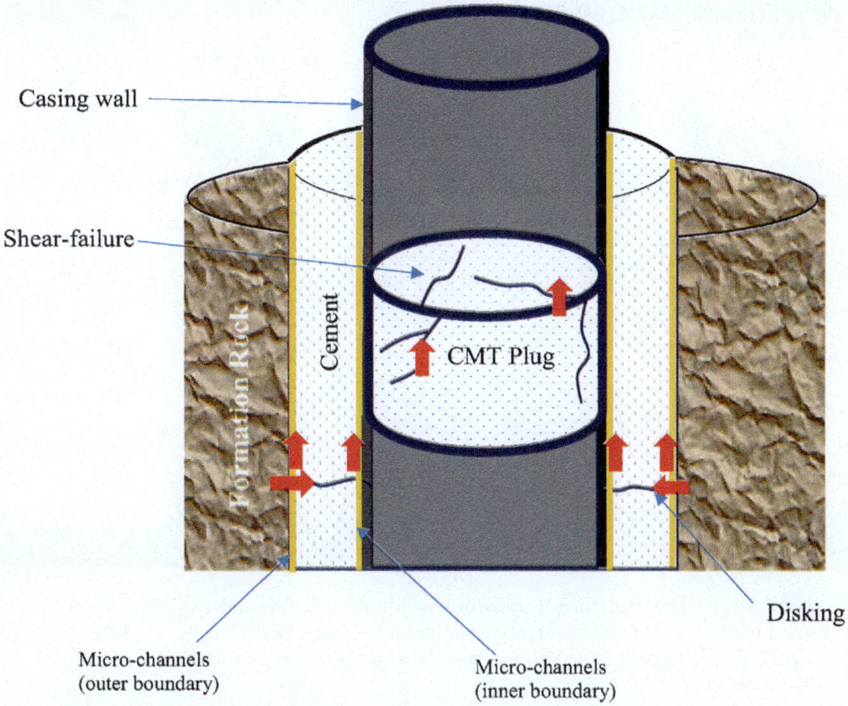

FIGURE 11.3 Different pathways of formation fluids through cements in a badly operated P&A well with different cement failure mechanisms shown.

well abandonment. The old, abandoned brine wells that had been drilled down into the Hutchinson Salt were only cased down through the shallow "Equus beds" aquifer. The deeper parts of the wells were open hole and provided paths for the gas to escape to the surface (Allison, 2001). Thus, after recognition of the problem source, up to 50 relief wells had to be drilled to control the situation.

11.3 Permanent plug and abandonment versus suspension

As the essential target of any P&A job, it is first important that proper sealing of different annuli in the wellbore is achieved (see Fig. 11.5). It is reminded from Section 11.4.3 that the A-annulus, also called "production casing annulus, PCA," refers to the space between the completion string and production casing; the B-annulus refers to the space between the production and intermediate casing, etc. Next, it was required to set a barrier across each reservoir zone (or flow zone) and some more barriers

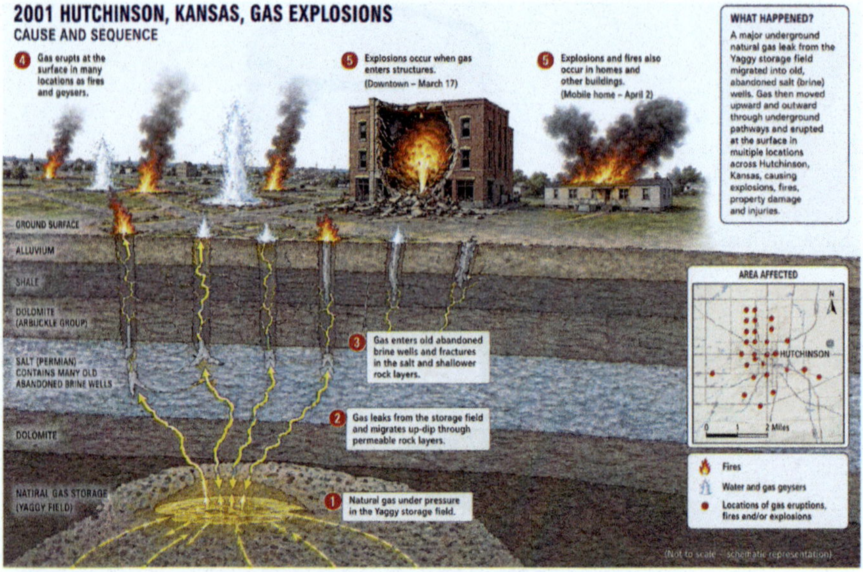

FIGURE 11.4 The Hutchinson blowout incident with fire and explosion occurring in Arkansas in 2001, due to bad well abandonment. *(Redrawn from Gosnell, N. (2021). Retrieved June 14, 2024, from https://salinapost.com/posts/c7bc68ec-4e72-4f7d-9d3e-07ddf962e049).*

above them in the borehole. Currently the caprock restoration criterion is the minimum MUST to be met; therefore the combination of caprock and annular cements should provide adequate hydraulic isolation and integrity. As the focus of the conventional P&A method was on shutting off all the reservoir zones, at least one barrier must be set in the well across each caprock (see Fig. 11.6A). However, recently, this focus is being modified and only a single barrier is sufficiently set at the topmost caprock (see Fig. 11.6B); thus new P&A standards are becoming less stringent and less costly.

There are two types of P&A: suspension and permanent. The suspension of a well refers to the temporary cessation of production or injection activities in a well, with the intention of maintaining the well in a state that allows for future operations, should they be needed. There are a variety of reasons for suspension, including preparation delay in construction of production pipeline or refinery construction or repair (if the well is planned to be produced), regulatory requirements or inspections, strategic decisions (e.g., waiting for market conditions to improve before abandoning the well), the need for further evaluation or data collection before final abandonment, and operational or technical issues that delay full abandonment.

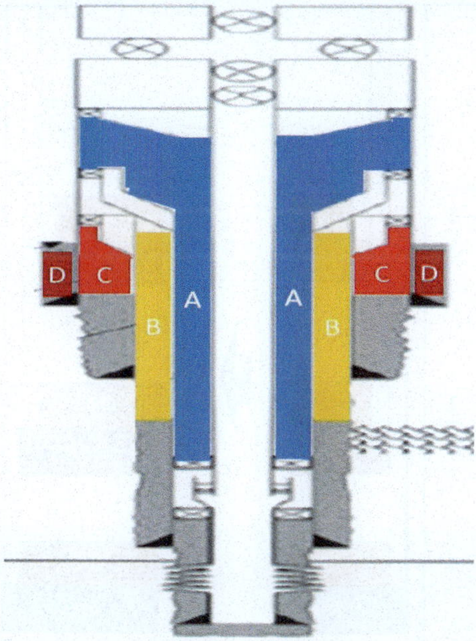

FIGURE 11.5 Different wellbore annuli named from A (between the completion string and production casing) to B (between production and intermediate casings), C, etc., and necessity of sealing them during production and following plug and abandonment if completion string is kept in the hole.

11.3.1 API standard

Based on API (2021), permanent abandonment barriers should include those placed:

- Across any exposed casing/liner shoe,
- In open hole,
- Above perforated intervals in cased holes,
- At points where casing has been removed,
- Across liner tops,
- Above and below usable water sources,
- Above or below hydrocarbon-bearing zones or other potential flow zones,
- At the surface or mudline.

Fig. 11.7 shows a schematic of permanent well abandonment requirement for barriers, showing the strict requirements asking for many cement plugs. Based on API (2021) the material used as the barriers or sealants (like plug 2 in Fig. 11.7 and plugs 1–4 in Fig. 11.6A) should exhibit almost

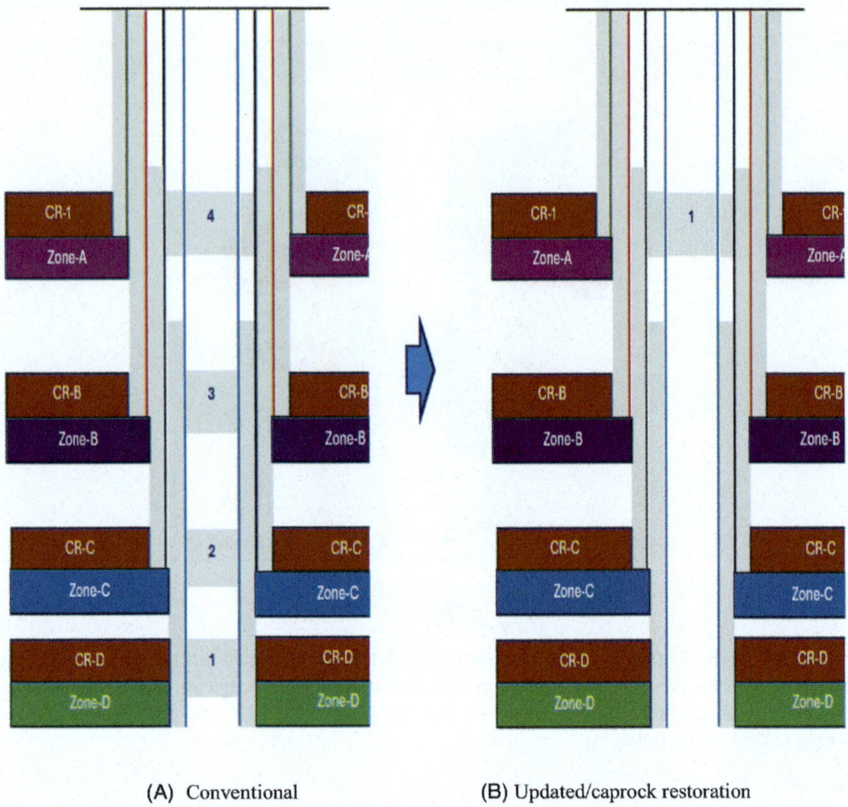

FIGURE 11.6 Minimum barriers, including plugs, needed for (A) conventional/ traditional plug and abandonment method (*left*) and (B) minimum barriers needed in modern (*right*) caprock restoration method in case of multiple reservoirs in a well.

the same properties as the cap rock, though it is nearly impossible for a sealant to be the same as the caprock. Therefore additional lab testing may be required to qualify existing or new materials for challenging chemical environments.

It is advised by API RP 65-3 that a mechanical base (e.g., a bridge plug) be set under the cement plug. "Mechanical plugs are typically used as a reliable base for cement that is placed above the mechanical plug. They aid in the placement of cement but are not typically considered a stand-alone permanent abandonment barrier." See Fig. 11.8 for an example schematic.

11.3.2 Operational procedure

In the permanent P&A, not only must poor annular cement intervals be repaired, but also the Christmas tree and uncemented portions of casing

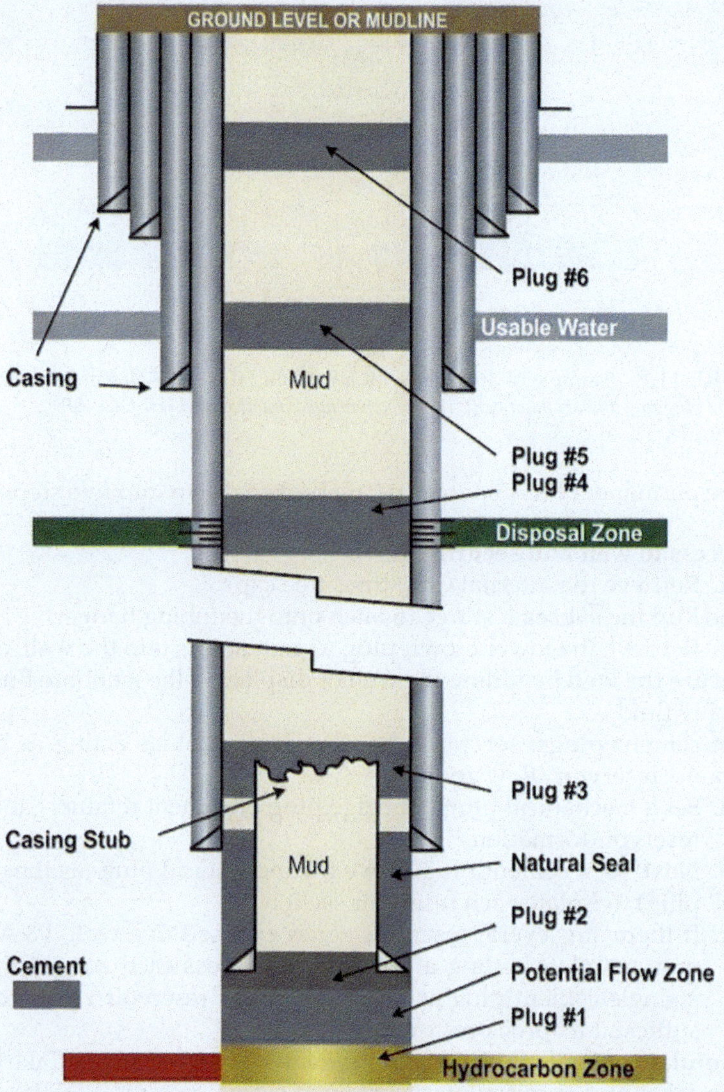

FIGURE 11.7 Example schematic of permanent well abandonment. *(Redrawn from API. (2021). Wellbore plugging and abandonment. API recommended practice 65-3 (1st Ed.). API).*

strings might also be removed depending on type of well design and construction. There are many challenges for P&A operations, such as missing documents of such old wells and serious corrosion of wellheads and infrastructure, making operations subject to possible incidents or blowouts. The procedure is briefed as follows:

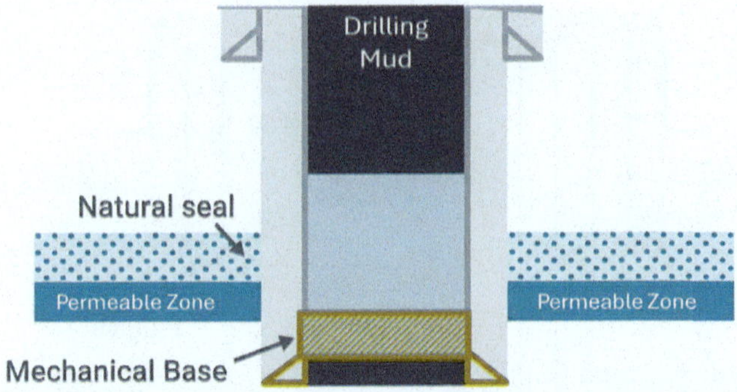

FIGURE 11.8 Example of a barrier with a mechanical base. *(Redrawn from API. (2021). Wellbore plugging and abandonment. API recommended practice 65-3 (1st Ed.). API).*

The permanent P&A operations are broken down into five steps:

1. **Access to well and securing it:**
   a. Retrieve the internal Christmas tree cap,
   b. Run the subsea test tree to latch onto the tubing hanger,
   c. Retrieve the lower crown plug to gain access into the well, and
2. **Secure the well** by killing the well or displacing the wellbore fluid with heavy fluid.
3. **Set cement plugs:** set/place barriers cement plugs acting as barriers against reservoir/flow zones:
   a. Set a mechanical plug (a bridge plug or cement retainer) above the reservoir formation.
   b. Next, set a cement plug above the mechanical plug (as the deepest plug) to isolate each reservoir section.
   c. If there are several reservoir zones exposed in a well, P&A policy may mandate setting a cement plug across each zone. Otherwise a single cement plug across the topmost reservoir zone would be sufficient (caprock restoration).
4. **Annular cement evaluation and remediation (if needed).** Take cement bond logs and monitor latest sustained casing pressures (SCPs). If SCPs were observed at the wellhead, remediation should be performed to isolate flow behind the casing. This involved remedial/squeeze cementing in free-pipe intervals (no cement).
5. **Cut and retrieve the completion/tubing string:** The completion string is cut using a downhole electric cutting tool above the production packer and retrieved.
6. **Cut and retrieve casing strings and wash cements behind:** Casing strings are not usually cemented up to the surface, though it depends on a company's policies. In this case, P&A standard usually requires

cutting and retrieving the casings up to the surface, as well as cleaning the cement behind them.

7. **Set top plug:** A surface/top or environmental plug should be spotted close to the seabed or near the surface; this is usually required for regulatory compliance.
8. **Clear and reclaim the location**, which involves the removal of subsea wellhead equipment to make the well site intact like before drilling. This job, which is called reclamation, is usually carried out using a diving support vessel postrig mobilization. This is to ensure that the location has been restored to its original state and can be considered for use as it was before drilling the well.

11.4 Evaluation

Upon P&A the primary barriers are the cement plug and the cement sheath isolating downhole formations, particularly the reservoir. The secondary well barriers are the backup barriers and are not in direct contact with the formation fluids. However, such barriers must be able to hold pressures in the case of failure of primary barriers. During production the casing string, cement, packer, wellhead, and Christmas tree provide the secondary barrier. Following P&A a second cement plug may be set above the primary barrier to form the secondary barrier. Sometimes, additional cement plug(s) may be set depending on company policy.

It is important to evaluate barriers carefully based on API, ISO, and NORSOK standards. For P&A, barrier evaluation should be conducted for long-term integrity.

11.4.1 Annular cement evaluation

Abandoning the well can be both intricate and expensive because of uncertainties surrounding the condition of the cement barrier in the B-annulus, which may have altered over the course of the well's lifespan. Conventional cement bond logs cannot operate well in double or multiple casing strings, which is the case with most well abandonment cases. Conventional cement bond logs cannot operate well in double or multiple casing strings, which is the case with most well abandonment cases. This lack was persistently existing for many years until recently some companies like SLB developed advanced technology tools to resolve the issue. A great example of such a technology is the epilogue tool, which is capable of evaluating cement bond even across multiple casing pipes where most conventional CBL tools fail to evaluate. See Chapter 5 (Section 5.5.1) for more information.

11.4.2 Cement plug evaluation

The evaluation methods for cement plugs have already been covered in Chapter 9, titled "Fundamentals of cement plugs and evaluation." Please refer to that chapter for a detailed discussion on the topic. To briefly summarize the common evaluation methods include the weight test, positive pressure test, and, in some special cases, the inflow test and negative pressure test.

11.4.3 Postabandonment monitoring and retrofitting

After decommissioning old wells, there are still safety and environmental concerns about their well integrity issues or subsequent leakage risk of methane or other fluids.

To prevent the associated risks and incidents like Hutchinson (Fig. 11.4), there are two solutions:

✓ Postabandonment monitoring: it may be required to verify the long-term integrity of the cement barrier and detect any potential leaks or failures over time.

✓ Retrofitting to other energies: the best way to ensure continuous monitoring of P&A wells is their retrofitting into geothermal or underground batteries. It is advantageous to convert or retrofit nonactivated or unprofitable oil and gas wells into geothermal wells in terms of deferring P&A costs (up to 40 years or longer), safety and environmental concerns, and potential energy production. Via retrofitting a repurposed geothermal well can stay active for a long time, and integrity of such wells can be monitored. It is a matter of fact that prior to retrofitting, possible well integrity issues of abandoned wells must be resolved; otherwise, the retrofitting scheme may fail, or the project life may be cut shorter, or there may be high associated costs and potential problems, for example, leakage of oil, gas, or water. In brief, this option is a more cost-effective option, as it contributes to not only eliminating monitoring costs but also bringing revenues from renewable geothermal energy production from abandoned wells. This signifies that we have transformed threat to opportunity (see Ashena (2023) for more information).

11.5 Creating annular barrier

In cases where the casing cement quality is inadequate or absent, it is necessary to remediate the cement by accessing the annular space behind the casing to install a reliable barrier inside and outside the casing. Traditionally, this target is achieved through a conventional method known

FIGURE 11.9 A section-mill assembly.

as section milling, which involves removing a portion of the casing using milling techniques. Specialized cutting tools are used to mill through casing steel, which is called *section milling*. Alternatively, newer methods such as perforate, wash, and cement (PWC) and melting downhole completion and casing have been developed to address this challenge.

11.5.1 Cut-and-pull and section-milling

In the section milling method a window is milled (above a casing coupling) to grind away a portion of casing and cement. If casings were not cemented to the surface (based on policy of some companies), the upper part of the casing string should be cut and pulled from the wellbore (cut-and-pull operation). Before cutting the casing a CBL operation is recommended to determine the top of cement (TOC) behind the TOC behind casing for efficient cutting and pulling operations. After TOC determination the casing is cut, usually at about 3.3 m (10 ft) above TOC. This increases the chance that the casing can be successfully pulled from the wellbore (see an example in Fig. 11.9). Obodozie et al. (2016) recommended using the high-expansion section milling tool; thus a full section milling operation of the 11 3/4 or 13 3/8-in casing would be carried out through an 8.5-in restriction. This enables establishing a rock-to-rock barrier across multiple annuli. As

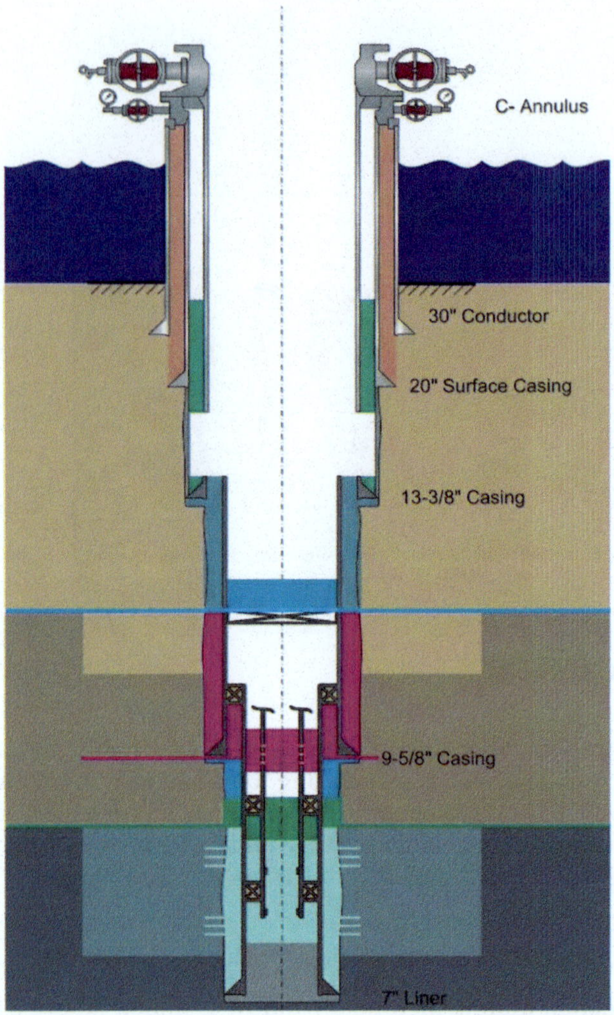

FIGURE 11.10 Case study well, 9 5/8-in cut, pulled and section milled 13 3/8-in casing. *(Redrawn from Obodozie, I. E., Trahan, S. J., & Joppe, L. C. (2016). Eliminating sustained casing pressure in well abandonment. In Paper presented at the Offshore Technology Conference Asia, Kuala Lumpur, Malaysia March 2016. https://doi.org/10.4043/26432-MS).*

advantages, this tool reduces cost and time of handling and disposal of materials from the wellbore during P&A.

During the milling section of the casing a lot of swarf (metal fillings) is produced. It is important that circulation is established and the hole is kept clean off swarf and other debris. Next the window is underreamed to expose new formation. Finally a cement plug is set across the window. Fig. 11.10 shows a photo of a section mill assembly.

11.5.2 Perforate, wash, and cement

The conventional section milling to remediate the cement integrity has been almost eliminated using PWC method, which has been widely used by Shell North Sea (Othman et al., 2018).

To carry out "PWC," first a scraper trip is essentially made to clean inside the casing, which is made by running the scraper assembly in the hole. Next a set of operations is conducted under the PWC. The PWC operations can be carried out either conventionally via three runs/trips (with each run for each step) via or two runs with all perforation-and-wash jobs being done in one run and cementing being conducted in the subsequent run. In the latter option, using advanced technology tools, the perforation gun and washing tool are run in a single run in the hole by which perforating is completed first; then an activation mechanism is engaged so that the gun is dropped in the rat hole of the well, and then the washing operation can be conducted. The washing operation is usually achieved via two substages: first washing by the wellbore/wash fluid and then by the spacer fluid (which is usually a water-based fluid). The two- or single-run operation is preferred over the three-run because of its customized time and cost. Fig. 11.11 shows that the work string has two sets of packers.

The procedure for the conventional PWC operations is as follows (see Fig. 11.11 for steps A–E and Fig. 11.12 for steps A–C):

1. **Perforation**
   a. Perforate the casing with the perforating gun either left in hole or retrieved.
2. **Conversion**
   a. Run in the hole with the washing tool downhole and start pumping the wash fluid to remove the debris (metal, mud filter cake, and settled materials) to clean the annular space behind the perforated casing. *Note*: The removed debris and metal can be observed on the shale shakers, which help to monitor the washing process.
   b. The washing operation continues downward until the tool reaches the bottommost perforations to clear off all debris across the formation.
   c. Start pumping the spacer fluid as the tool is pulled upward.
   d. With the tool positions near the top perforations the spacer is pushed further up beyond the uppermost perforations in the casing.
3. **Cementing**
   a. The bottom hole assembly (BHA) is placed below the bottom of perforations, and they start pumping some volume (a few barrels) of cement.

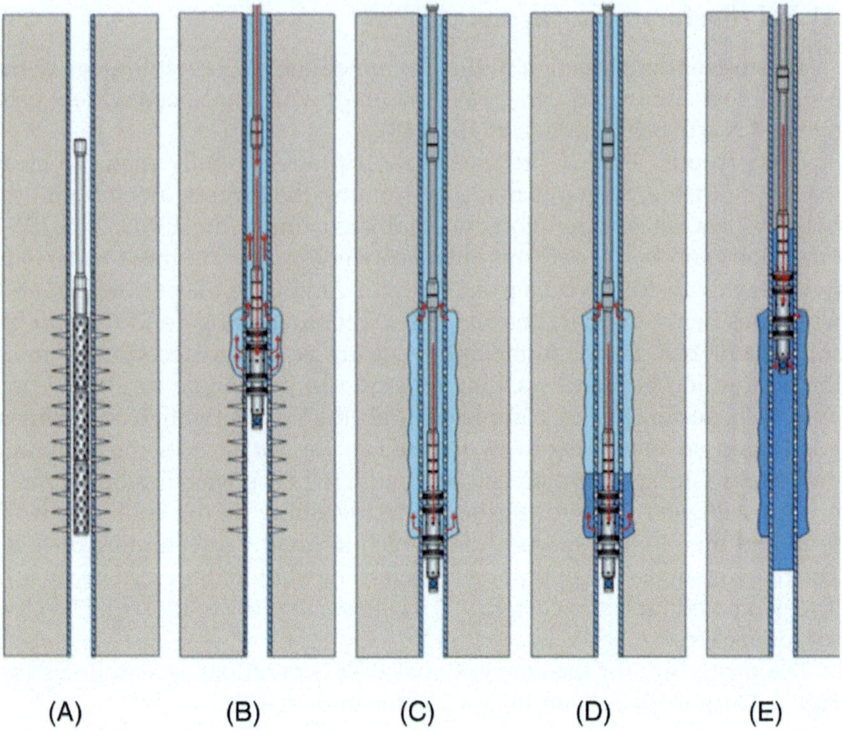

FIGURE 11.11 The perforate, wash, and cement technique: (A) First the casing is perforated. (B) A washing tool is run in the well above perforations. Start pumping wash fluid to clean the annulus behind the casing while moving the tool downward. (C) The bottom hole assembly of the tool is positioned below the lowermost perforations in the casing. (D) The spacer fluid is pumped, and the work string is pulled upwards; and (E) the spacer is pushed further up beyond the uppermost perforations in the casing (Khalifeh & Saasen, 2020; Archer Oiltools).

b. Pump-and-pull while cementing, that is, pump the cement slurry as the work string is pulled out.
c. Pump cement and circulate out the cement in BHA, pull the BHA out of cement, at least two stands above the TOC.

Note that there are two available PWC systems: jet-type and cup-type. The jet-type system, shown in Fig. 11.13A, applies jets to perform the washing, where the jets were positions with differing angles to allow the greatest potential of cleaning the annulus. However, the cup-type PWC system, shown in Fig. 11.13B employs a collection of flow ports created from the perforations, separated by sealing cups, to force fluid out through a perforated interval, with the purpose of circulating out any possible debris contained in the annulus. This system was used for the first subsea

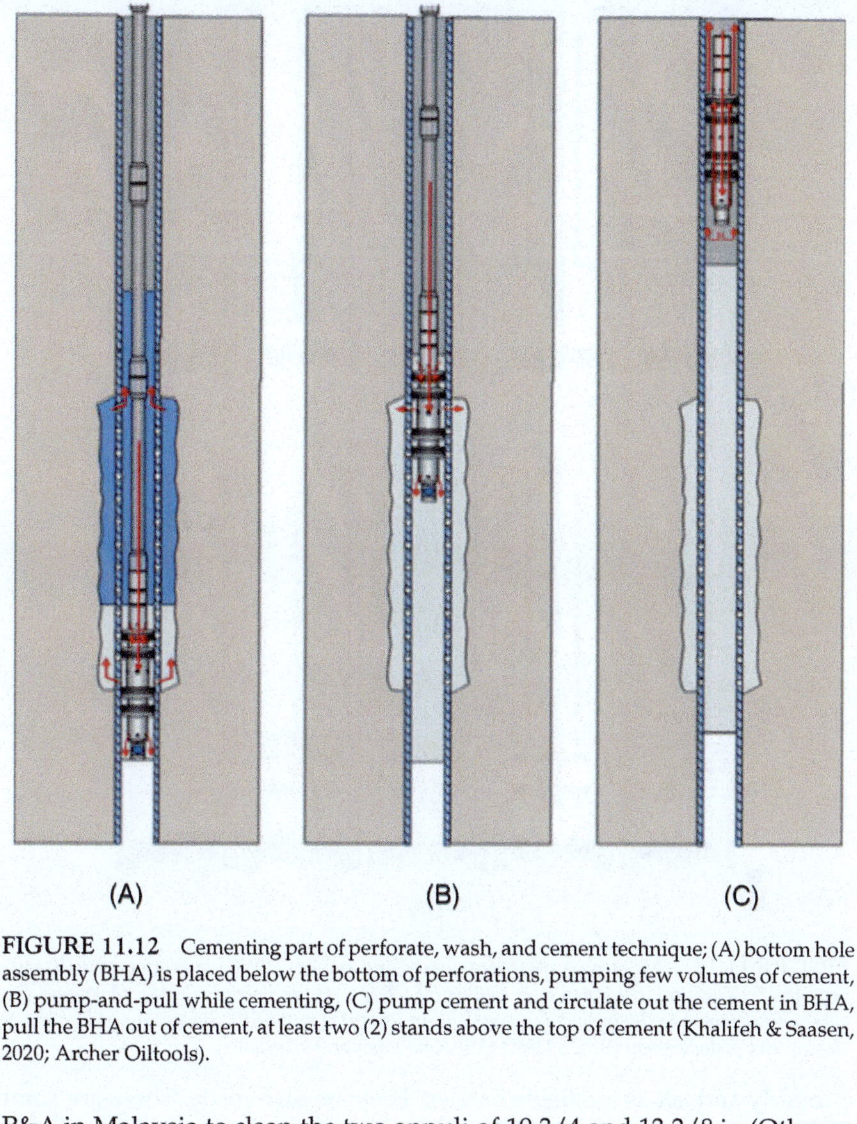

FIGURE 11.12 Cementing part of perforate, wash, and cement technique; (A) bottom hole assembly (BHA) is placed below the bottom of perforations, pumping few volumes of cement, (B) pump-and-pull while cementing, (C) pump cement and circulate out the cement in BHA, pull the BHA out of cement, at least two (2) stands above the top of cement (Khalifeh & Saasen, 2020; Archer Oiltools).

P&A in Malaysia to clean the two annuli of 10 3/4 and 13 3/8 in (Othman et al., 2018).

11.5.3 Melting bismuth and termite

Cements, when it fails, cannot block gas from sealing properly. Bridge plugs lack great enough expansion ratio required in some wells and rely on elastomers to seal, limiting their ability to seal in damaged casing, tubing, or open-hole sections. After being injected, resins need time to seal

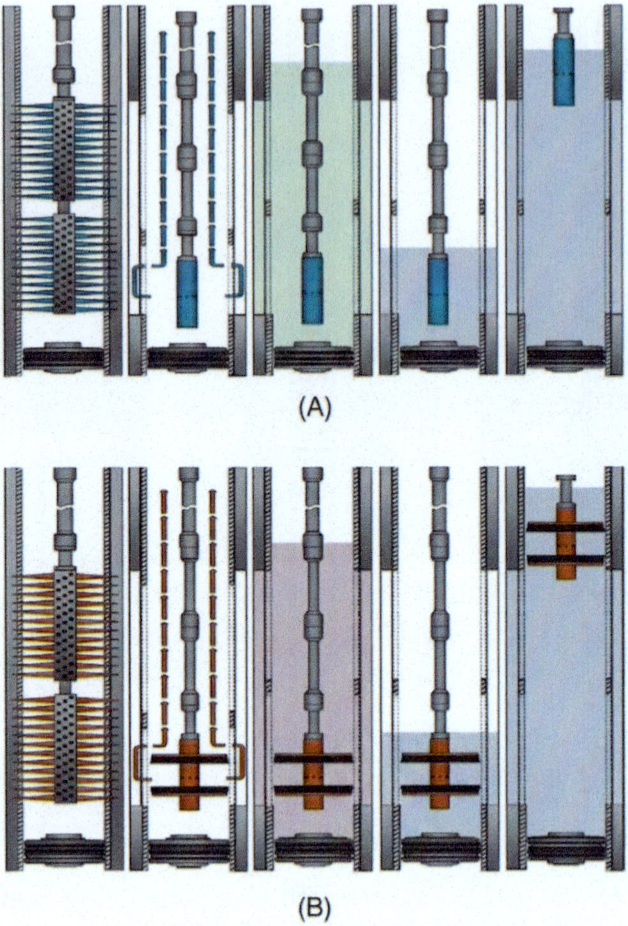

FIGURE 11.13 (A) Jet-type perforate, wash, and cement (PWC) operation and (B) cup-type PWC operation. *(Redrawn from Othman, F., Teo, C. H., & Bakri, F. (2018). Malaysia's first subsea development well plug and abandonment. In Presented at the SPE Symposium of Decommissioning and Abandonment (SPE-193980-MS), Kula Lumpur, Malaysia).*

properly and act as a reliable barrier. They are also costly. These are some limitations of the current sealing practices by the industry. To overcome the limitations the latest technology is called bismuth and termite (B&T), which was recently developed to replace the traditional cement plugging method and bridge plug for P&A.

B&T is a type of metal-based fuel in the form of thermite that can be used in well P&A operations. Bismuth alloys generally have lower melting points than traditional materials like lead or other heavy metals. This makes them easier to work with when sealing wells, as the material can melt and flow more easily while being pumped. The term "thermite"

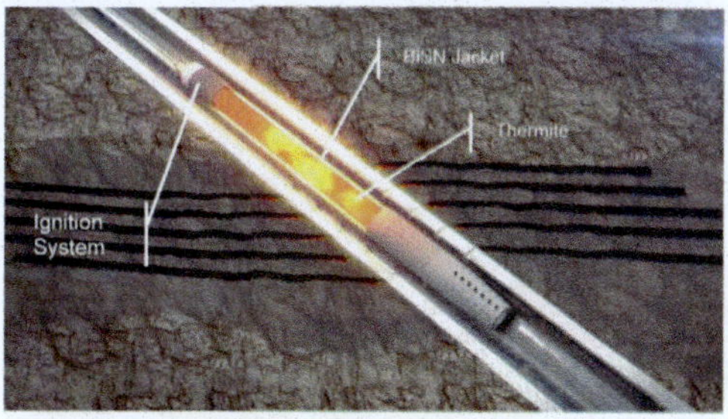

FIGURE 11.14 The bismuth and termite (B&T) method for barrier creation in the well. *(Redrawn from Carragher, P., & Fulks, J. (2018). Well abandonment solutions utilizing bismuth and thermite. In Paper presented at the Abu Dhabi International Petroleum Exhibition & Conference, Abu Dhabi, UAE. https://doi.org/10.2118/193118-MS).*

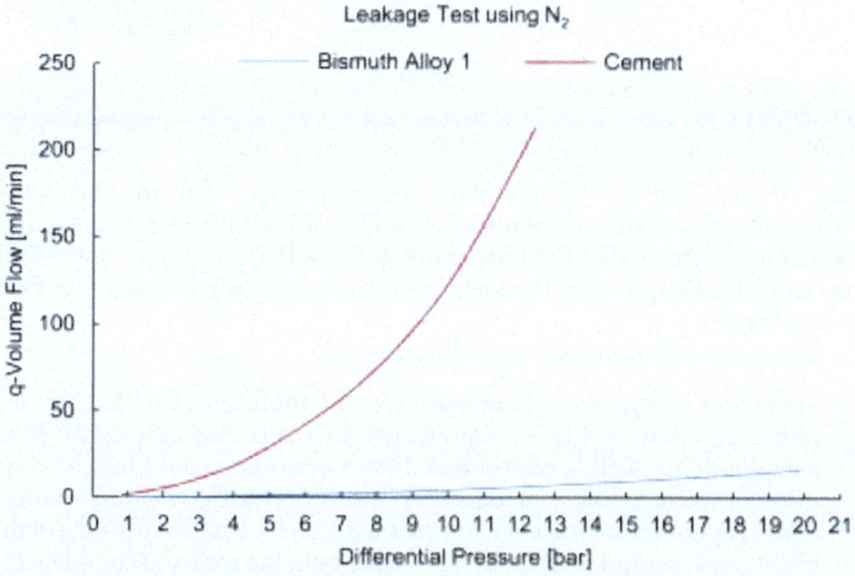

FIGURE 11.15 Bismuth and Termite (B&T)'s method of sealing compared to bridge plug and cement in the case study. *(Redrawn from Carragher, P., & Fulks, J. (2018). Well abandonment solutions utilizing bismuth and thermite. In Paper presented at the Abu Dhabi International Petroleum Exhibition & Conference, Abu Dhabi, UAE. https://doi.org/10.2118/193118-MS).*

refers to a mixture of metal powder (often aluminum) and metal oxide (such as iron oxide), which burns at extremely high temperatures. This is particularly useful in creating an intense localized heat that can melt or break down wellbore structures/materials (casings and formations), enabling rapid and effective plugging of wells, effectively creating a seal and contributing to well integrity. This is particularly useful in operations where time is of the essence and quick abandonment is needed.

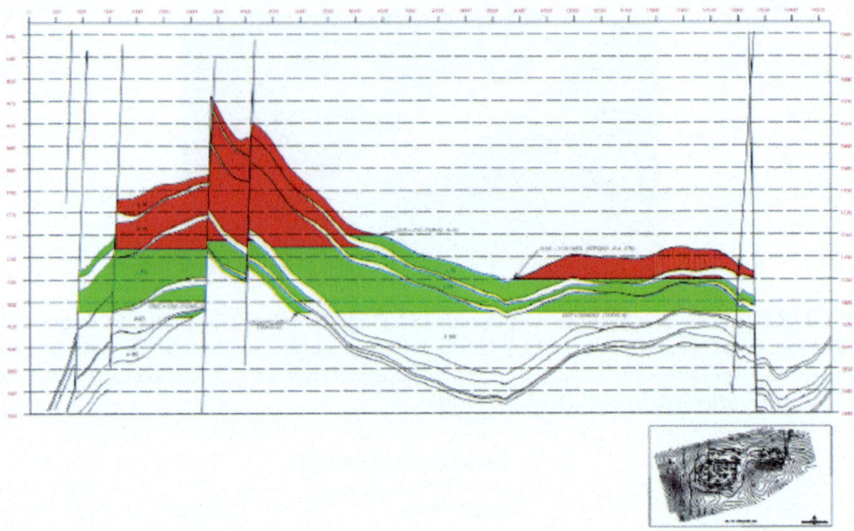

FIGURE 11.16 Cross-section of the depleted reservoir for permanent plug and abandonment.

The combination of both materials contributes to enhancing the safety, efficiency, and environmental responsibility of well P&A operations. It is worth mentioning that the idea is not new; rather, the patent dates back to the 1930s. Despite this, the method was not seriously practiced for P&A until recently.

The said method has several advantages:

- ✓ Time- and cost-saving advantages over the traditional method because of its simplicity and less requirements. In a specified case study of an average depth well, a conventional P&A operation could take 12 days to be completed, which included perforation as well as wash and cement (PWC) operations. However, the B&T method took half of the time of the traditional method (6 days) to set two plugs in the well and complete the plugging job.
- ✓ As another advantage of the latter method, it is pumpable, as the material has almost the viscosity of water when in liquid phase and it becomes very integrated when it solidifies. Bismuth alloys, when cooled and solidified, are known for their strength and resistance to corrosion. This is beneficial for long-term sealing of wells, where the plug must remain intact for many years or even decades. Therefore, in the case of using B&T for P&A, there is no need for auxiliary jobs such as borehole cleaning, cement plug tagging, etc.
- ✓ B&T offers a safe, environmentally friendly, low-melting, and durable solution for creating long-lasting seals in well abandonment. It is a quick

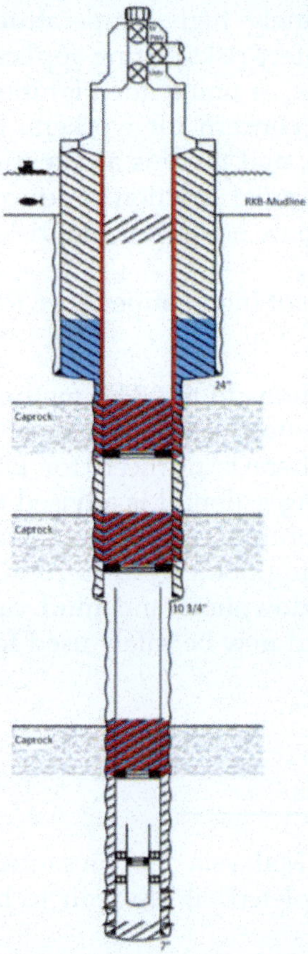

Primary barrier elements		
Element	**Qualification**	**Monitoring**
Cement plug #1	Pressure test to 1000 Psi	Pressure above plug
Cement plug #1 - Annulus side	Pressure test to 1000 Psi	Pressure above plug
Cement plug #2	Pressure test to 700 Psi	Pressure above plug
Cement plug #2 - Annulus side	Pressure test to 700 Psi	Pressure above plug
Cement plug #3	Pressure test to 700 Psi	Pressure above plug
Cement plug #3 - Annulus side	Pressure test to 700 Psi	Pressure above plug
Cement plug #3 - Annulus side	Pressure test to 700 Psi	Pressure above plug
Secondary barrier elements		
Element	**Qualification**	**Monitoring**
Cement plug #1	Pressure test to 1000 Psi	Pressure above plug
Cement plug #1 - Annulus side	Pressure test to 1000 Psi	Pressure above plug
Cement plug #2	Pressure test to 700 Psi	Pressure above plug
Cement plug #2 - Annulus side	Pressure test to 700 Psi	Pressure above plug
Cement plug #3	Pressure test to 700 Psi	Pressure above plug
Cement plug #3 - Annulus side	Pressure test to 700 Psi	Pressure above plug
Surface casing	Pressure test to 500 Psi	C-annulus pressure

FIGURE 11.17 Final well schematic at the end of plug and abandonment.

and effective method, particularly in hazardous workover well cases where wellbore integrity has been severely compromised and a rapid solution is needed.

Although the B&T method offers several advantages in well P&A, there are also some disadvantages to consider:

✓ Bismuth: Because of their limited availability, bismuth alloys can be more expensive compared with traditional plugging materials like cement. Next, bismuth alloys can be more brittle compared to other plugging materials, especially when subjected to high temperatures and mechanical stresses. This could lead to cracking or failure of the seal under certain conditions.

✓ Termite: Thermite reactions generate extremely high temperatures (>2500°C or 4500°F), which pose significant safety risks during application. During deployment the risk of fire, burns, or unintended ignition of nearby materials can be a serious safety concern for workers. In addition, it is difficult to control the reaction, and there is a potential for wellbore casing damage. Finally the thermite reaction produces metal oxides and other byproducts that may pose environmental risks, especially if not carefully managed.
✓ Displacement is a potential challenge because of high temperature following the reaction.

To overcome the drawbacks, recent modifications to the B&T method in well P&A have focused on improving the effectiveness, safety, and environmental sustainability of the process. There is no need for DH electrical heaters (480 V and 11 A) for the melting action. It is advised to use modified materials and damping agent to control huge temperatures and reduce 2000°C temperature (to reduce melting point to 93°C–263°C) such that it would not be hazardous for the downhole and mud can also be pumped. Therefore even the drilling mud may be safely used for displacement (Figs. 11.14 and 11.15).

11.6 Case study wells

In this part, two case studies of P&A in Malaysia are presented. Fig. 11.16 shows a cross-sectional view of a depleted oil reservoir to be abandoned.

11.6.1 Case study I

Fig. 11.17 shows four cement plugs set in the well following the permanent P&A method. Fig. 11.18 provides a more specific display of the process of well abandonment and where each cement plug was set across each cap rocks. This method was followed because the operator's strategy was conventional, that is, dictating setting at least one cement plug across each cap rock formation. The shallowest plug is called the environmental or surface plug, which is not considered a barrier.

Fig. 11.19 shows well schematics before and after the well abandonment using the caprock restoration method, which involves using only a single cement barrier plug across the shallowest caprock. As can be seen the completion string has not been removed from the wellbore, but rather it has been maintained in its place. Therefore the workover operations can be performed with a hydraulic control unit rather than a rig. There would be no requirement for using a drilling or workover rig to lift the

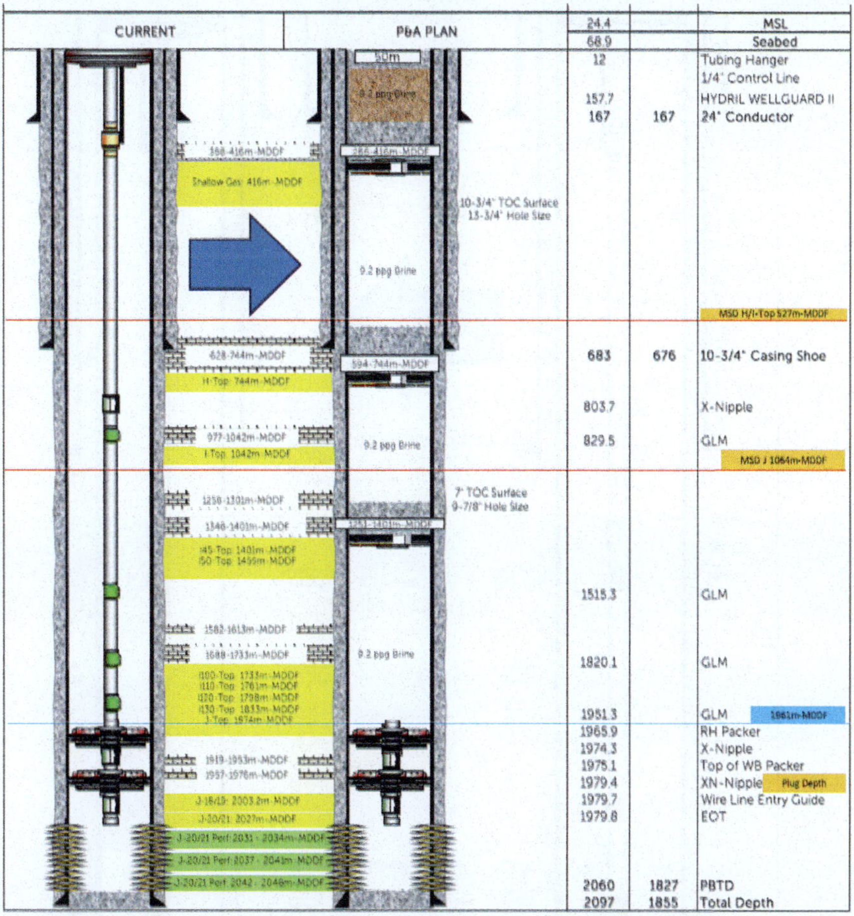

FIGURE 11.18 Well schematics before (*left*) and after plug and abandonment (*right*) with the traditional method using a cement layer across each reservoir zone or caprock.

completion string and pull it out, which would cut operations cost. As less cement plugs are required in the recent method, several runs (of running in and pulling out of hole) to set cement plugs can be eliminated via this method. In brief the advantages of following the criterion of caprock restoration method for well abandonment are obviously cost and time minimization.

11.6.2 Case study II

Otman et al. (2018) presented a subsea P&A case study offshore Sarawak in Malaysia. The wells were producing hydrocarbons from 1999 to 2015

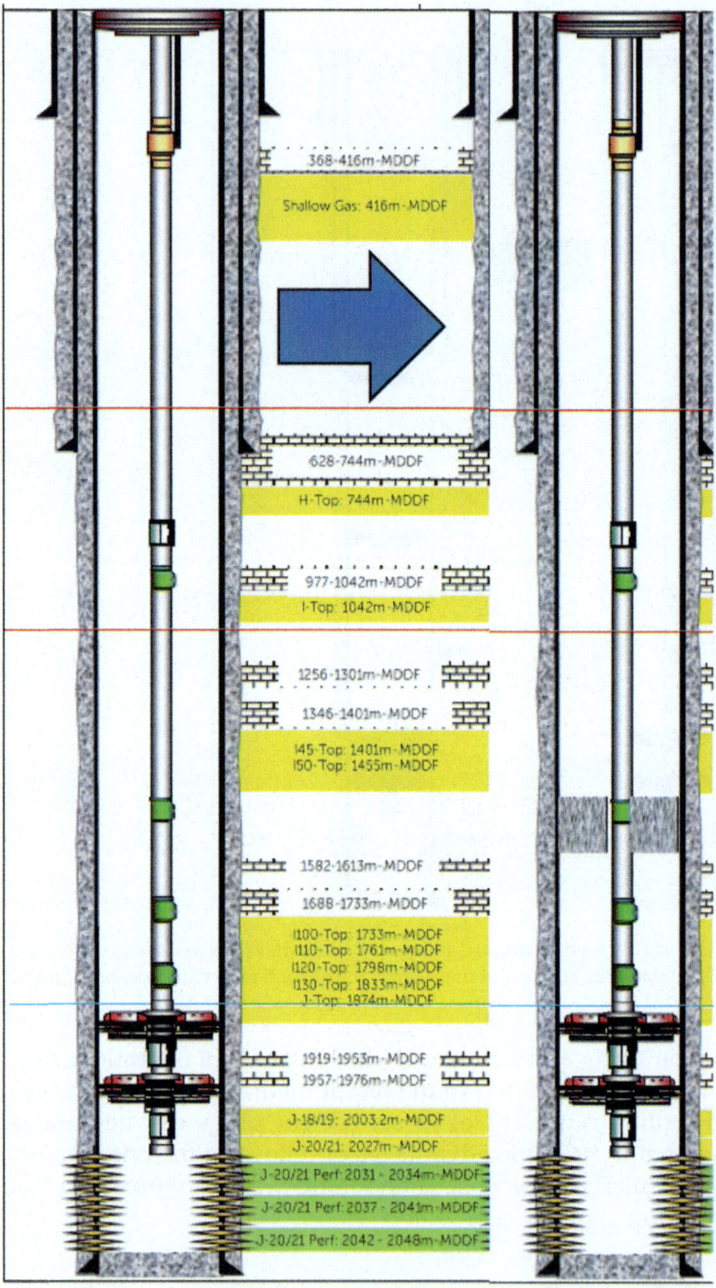

FIGURE 11.19 Well schematics before (*left*) and after plug and abandonment (*right*) with the recent caprock restoration method using a single barrier above the caprock formation.

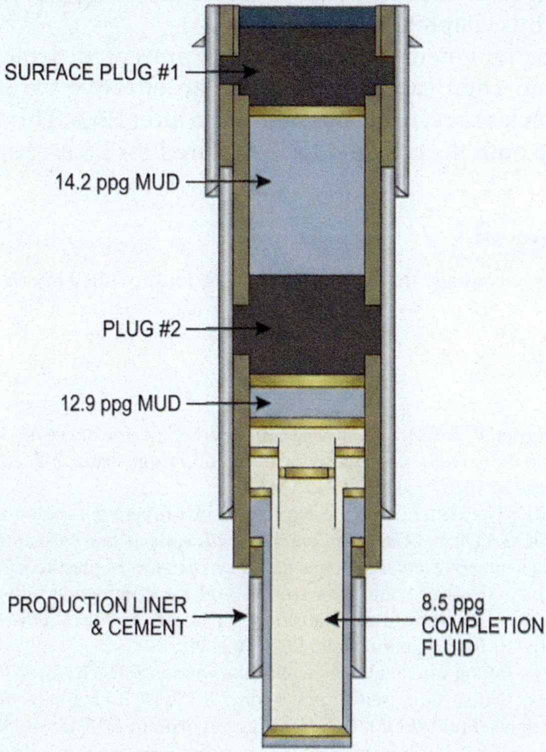

FIGURE 11.20 A subsea plug and abandonment case study. *(Redrawn from Othman, F., Teo, C. H., & Bakri, F. (2018). Malaysia's first subsea development well plug and abandonment. In Presented at the SPE Symposium of Decommissioning and Abandonment (SPE-193980-MS), Kula Lumpur, Malaysia).*

when the production limit was reached in the field, and P&A was programmed. This P&A campaign did not require retrieval of any casing, wellhead, or subsea equipment (Fig. 11.20).

11.7 Summary

1. The fundamental principles of cement plugs and well P&A were explained to ensure that P&A operations are conducted according to safe standards. Postabandonment monitoring was discussed as an essential means of evaluating well integrity after P&A.
2. Various methods of creating annular barriers were covered. Alternative P&A techniques, such as B&T, were explored as potential rivals to caprock restoration using cement plugs.

3. Different types of cement plugs used for restoring well integrity were detailed in this chapter.
4. Two well abandonment case studies were presented, demonstrating that caprock restoration can be considered an effective barrier to prevent hydrocarbon leakage, flow, and migration after P&A. This method helps to minimize both the cost and time required for P&A operations.

Acknowledgment

The author wishes to cordially thank Mr. Razak Yakob for providing information for the first case study.

References

Abdelal, G., Carragher, P., & Clark, B., Investigating creep performance of Wel-lok BiSN plug for sealing of offshore wells, Geoenergy Science and Engineering, 242, 2024, 213153, https://www.doi.org/10.1016/j.geoen.2024.213153.

Allison, M. L. (2001). The Hutchinson gas explosions: Unraveling a geologic mystery. In *26th Annual KBA/KIOGA Oil and Gas Law Conference: v1*. Kansas Bar Association 3-1–3-29.

API. (2021). *Wellbore plugging and abandonment. API recommended practice 65-3* (1st ed.). API.

Ashena, R. (2023). Analysis of some case studies and a recommended idea for geothermal energy production from retrofitted abandoned oil and gas wells. *Geothermics, 108*, 102634. https://doi.org/10.1016/j.geothermics.2022.102634.

Dragland, Å. (2024). Taking care of elderly wells. Norwegian SciTech News. Retrieved June 18, 2024, from https://norwegianscitechnews.com/2015/11/taking-care-of-elderly-wells/.

Environmental Defense Fund (EDF). (2023). Plugging orphan wells across the United States. Retrieved November 9, 2023, from https://www.edf.org/orphanwellmap.

Groom, M. (2020). Special report: Millions of abandoned oil wells are leaking methane, a climate menace. Reuters. Retrieved June 18, 2024, from https://www.reuters.com/article/us-usa-drilling-abandoned-specialreport-idUSKBN23N1NL.

Hmadeh, L., Jaculli, M.J., Elahifar, B., & Sangesland, S., Development of bismuth-based solutions for well plugging and abandonment: A review, Petroleum Research, 9(2), 2024, 250-264, https://www.doi.org/10.1016/j.ptlrs.2024.01.003.

Khalifeh, M., & Saasen, A. (2020). *Introduction to permanent plug and abandonment of wells*. Springer International Publishing https://doi.org/10.1007/978-3-030-39970-2.

Liversidge, D., Taoutaou, S., & Agarwal, S. (2006). Permanent plug and abandonment solution for the North Sea. In *Paper presented at the SPE Asia Pacific Oil & Gas Conference and Exhibition, Adelaide, Australia* September 2006 https://doi.org/10.2118/100771-MS.

NBC. (2023). Millions in the U.S. live near abandoned oil and gas wells linked to explosions and toxins. Retrieved November 9, 2023, from https://www.nbc.com/nbc-nightly-news/video/millions-in-the-us-live-near-abandoned-oil-and-gas-wells-linked-to-explosions-and-toxins/NBCN986309805#:⟨:text=on%20the%20app.-,Millions%20in%20the%20U.S.%20live%20near%20abandoned%20oil%20and,linked%20to%20explosions%20and%20toxins&text=Abandoned%20oil%20and%20gas%20wells%20can%20cause%20toxins%20and%20dangerous,are%20long%20out%20of%20business.

Obodozie, I. E., Trahan, S. J., & Joppe, L. C. (2016). Eliminating sustained casing pressure in well abandonment. In *Paper presented at the Offshore Technology Conference Asia, Kuala Lumpur, Malaysia* March 2016 https://doi.org/10.4043/26432-MS.

Osundare, O. S., Teodoriu, C., Falcone, G., & Ichim, A. (2018). Estimation of plugging and abandonment costs based on different EU Regulations with application to geothermal wells. In *Stanford Geothermal Workshop* https://pangea.stanford.edu/ERE/db/IGAstandard/record_detail.php?id=28266.

Othman, F., Teo, C. H., & Bakri, F. (2018). Malaysia's first subsea development well plug and abandonment. In *Presented at the SPE Symposium of Decommissioning and Abandonment (SPE-193980-MS), Kula Lumpur, Malaysia*, https://doi.org/10.2118/193980-MS.

Washington Post (2024). The U.S. has more than 120,000 abandoned oil wells, research shows. The Washington Post. Retrieved June 18, 2024, from https://www.washingtonpost.com/politics/2022/12/02/us-has-more-than-120000-abandoned-oil-wells-research-shows/.

CHAPTER

12

Cement integrity in carbon capture, utilization, and storage wells

Rahman Ashena[a], *Ibnelwaleed A. Hossein*[b], *and Asad Elmgerbi*[c]

[a]Smart Energy Data Analytics (SEDA)-Group, Houston, TX, United States, [b]Qatar University, Doha, Qatar, [c]Montanuniversität Leoben, Leoben, Austria

12.1 Introduction

The energy sector worldwide has been focusing on utilizing carbon dioxide (CO_2) to mitigate the adverse impact of global warming and decrease CO_2 emissions footprint. Field applications revealed the simultaneous advantage of injecting CO_2 into depleted oil reservoirs to improve the oil recovery and sequester CO_2. Even though the process may seem simple at first glance, carbon sequestration is far from immediate. Its effectiveness depends on long-term dynamics, meaning it primarily impacts future CO_2 levels rather than current ones. In the early stages of implementation, many critical questions remain unanswered, for example, HOW much CO_2 can actually be stored? How much of it dissolves? How fast does it migrate through different rock formations and the cement sheath? And how much of the surrounding rock is altered through dissolution and reprecipitation caused by chemical reactions?

Considering the CO_2-enhanced oil recovery (EOR) process the impact of CO_2 on cement integrity is significant for both injection and production wells. In this process, cement will be continually subjected to thermal and mechanical loading as well as carbonation after long contact with the injected supercritical CO_2. These stresses might significantly damage the cement sheath, creating microchannels, cracks, and pathways between

Well Cement Integrity
DOI: https://doi.org/10.1016/B978-0-443-51564-4.00012-6

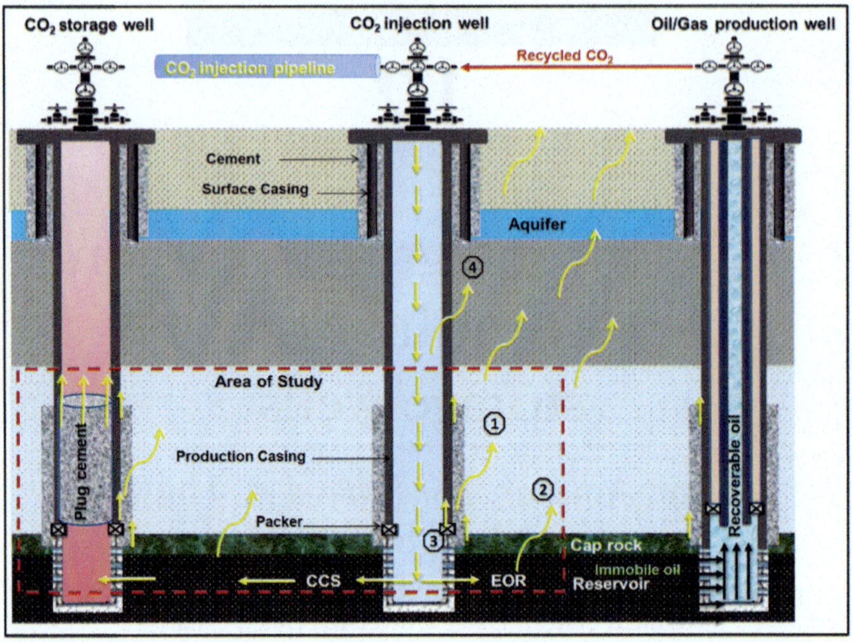

FIGURE 12.1 CO_2 emission mechanisms in an injection well for enhanced oil recovery (EOR) and carbon capture storage (CCS) applications (Geraci et al., 2017).

the cement–casing–formation interfaces and the bulk cement. The leak intensity is primarily dependent on the cement's permeability, its capillary properties, and the quality of the interfacial bonding between the casing, cement, and formation interfaces (Carey et al., 2007; Kiran et al., 2017). In the case of loss of cement integrity, remedial cementing jobs are executed to seal the overwhelmed cement, maintain the cement integrity, and avoid leakage of CO_2 to the surrounding environment; however, these jobs are considerably costly and may not yield good results.

This chapter aims to discuss CO_2-resistant cement systems and evaluation methods to sustain cement integrity under a corrosive downhole environment.

Fig. 12.1 illustrates a conceptual model of a CO_2-EOR system. It depicts the injection of recycled CO_2 through dedicated pipelines into subsurface geological formations containing residual oil, with production wells capturing the mobilized hydrocarbons. The schematic highlights key features, including caprock for containment, production zones, and potential fluid migration pathways. In alignment with the Geraci et al. (2017) report, this visualization highlights the importance of well integrity, monitoring infrastructure, and regulatory oversight in preventing potential risks, including CO_2 leakage associated with long-term underground injection in EOR operations.

12.2 Cement integrity challenges in carbon capture, utilization, and storage wells

In carbon capture, utilization, and storage (CCUS) the long-term performance of cement is a critical issue. Cement bonds need to withstand decades of CO_2 injection and pressure cycles without degrading or failing. The risk of failure in the cement sheath can be thought of for the following reasons:

1. When CO_2 is combined with water, it can form carbonic acid. The conventional Portland cement is susceptible to dissolution and degradation when exposed to acidic carbonic acid, especially in the long term, which could lead to failures or weakening of the cement bond, resulting in compromised well integrity.
2. CCUS wells are often drilled to significant depths where temperature and pressure can be much higher than in conventional oil and gas wells. This can affect the performance of the cement, leading to issues like poor bonding or cement cracking (see Chapter 13 for more information).
3. Because of sudden and significant temperature drops occurring because of supercritical CO_2 injections, huge thermal stresses can be induced in the cement, which could cause cracking or shrinkage, further weakening the cement bond and potentially allowing fluid migration.
4. Injection of supercritical CO_2 into salt formations may lead to salt precipitation near the wellbore, eventually forming salt crystals that can expand, exert mechanical stress on the cement sheath, and potentially cause failure.
5. CO_2 might stay trapped underground for a long time before anything happens. But that gives it more time to slowly react with cement in hidden ways—damage might not show up until much later.

12.3 Ordinary Portland cements and integrity issues

Ordinary Portland cement (OPC) is mainly composed of calcium (Group 2 of the periodic table) and silicon (Group 4) elements and has been the dominant material for cementing, most likely because of its known chemistry, affordability, and availability (Nelson & Guillot, 2006). However, there are multiple issues with using OPCs for CCUS. We categorize the issues into those related to the cement production phase and the operational ones associated with the mechanical sealing or integrity capability of cement.

The production-associated issues are:

- Production of OPC is not environmentally safe because its production gives out huge CO_2, and it is the third largest source of CO_2 emission globally. The current contribution of greenhouse gas emissions from OPC production is about 1.35 billion tons annually, or about 7% of the total greenhouse gas emissions to the Earth's atmosphere. The production of Portland cement worldwide is increasing by 3% annually. Around two-thirds of CO_2 emissions from cement production come from chemical process emissions, with the rest energy related. Through the production of 1 t of OPC, decarbonation of limestone occurs in the kiln, which is responsible for the liberation of 1 t of CO_2 to the atmosphere, as shown in reaction Eq. (12.1).

$$CaCO_3 + 2SiO_2 \rightarrow 3CaO \bullet SiO_2 + 2CaO \bullet SiO_2 + 5CO_2 \tag{12.1}$$

- Therefore Portland cement is mainly composed of calcium silicate ($CaO \bullet SiO_2$), with some calcium aluminate ($CaO \bullet$ and Al_2O_3) and ferrite/iron oxide (Fe_2O_3).
- Therefore Portland cement is mainly composed of calcium silicate ($CaO \bullet SiO_2$), with some calcium aluminate ($CaO \bullet$ and Al_2O_3) and ferrite/iron oxide (Fe_2O_3). The OPC is among the most energy-intensive construction materials, after aluminum and steel. The manufacturing of Portland cement requires considerable energy consumption for clinker calcination (which occurs at ~1450°C) and for the grinding of raw materials and clinker. The heat consumption is estimated to be about 3.8 GJ/t clinker. Such high energy requirements are associated with high gas emissions.
- Manufacturing of OPC deals with a potential future shortage of low-cost raw materials. As an example, in 2011, the world cement production was around 3.6 billion tons, requiring more than 3 billion tons of limestone. In China, there are currently 1326 limestone quarries, containing approximately 56.12 billion tons of limestone. Considering future growth in demand for cement, it is estimated that these deposits can meet the need for cement manufacturing for only the next 60 years.

The next category of OPC's issues is operational and is associated with their mechanical and sealing characteristics in wells; some extreme cases are mentioned below:

- OPCs may not sustainably withstand high mechanical stresses because of high injection pressures and temperature conditions and thus undergo mechanical failure, leading to matrix fractures or debonding from the casing or formation at the interfaces.
- OPCs may not withstand high chemical stresses because of corrosive environments, such as CO_2 or saline brine, and thereby undergo chemical degradation or corrosion, causing loss of cement integrity either in the matrix or at interfaces. This problem is the most extreme in corrosive environments such as CCUS wells or wells with formations containing high CO_2 or H_2S contents.

- In the long term, OPC's integrity may be lost either in the matrix or interfaces during the production phase or following the abandonment.

To overcome the OPC's operational issues, there are two common approaches: either using additives to improve OPC's mechanical and chemical properties or replacing OPCs with alternative cements. With the first approach, numerous works have been carried out, but with limited success for the industry. As for the second approach, various technologies can be applied to produce non-Portland cements, including partially prehydrated C–S–H binders, belite-rich cements, magnesium oxy-carbonate cements, magnesium silicate cements, calcium carbonate cements, and geopolymer cements (GPCs). Because of geopolymers' environmental advantages, this work focuses on developing GPCs for special challenging wells such as CCUS well cementing, which has been raised as an option by some researchers, but still it is mainly confined to research context or small-scale pilot tests and faces with several material science and operational uncertainties.

12.4 Additives and improved integrity

Multiple potential additives, comprising nanooxides and polymers, can be used to assess their ability to improve corrosion resistivity within the cement matrix. Such additives can expand to compensate for cement shrinkage, and thus possible flow channels and fluid migration may significantly decrease. The suggested cement formulations can be implemented in primary and remedial cement jobs to seal fractures and remedy the leakage pathways between the cement–casing–formation or leak paths through the set cement. The short- and long-term properties of the developed cement slurries should be evaluated, considering API indicators, to prevent CO_2 leakage. In the experimental approach to finding the best additive the rheological characteristics, such as apparent viscosity, plastic viscosity, thickening, and elastic behavior, as well as the mechanical properties, should be measured and compared. The mechanical properties should be found through nondestructive and destructive tests to obtain compressive modulus and Poisson's ratio. Because of susceptibility of OPCs, there are challenges in using this approach.

Some corrosion-resistant nanomaterials are nanooxides such as Al_2O_3, SiO_2 (nanosilica), and Fe_2O_3 (Ahmed et al., 2018; Jafariesfad et al., 2020), TiO_2 (Zhang & Bachu, 2011), and copper oxide (Isahak et al., 2013). Among them, nanosilica proved to be a good choice for improving pozzolanic reactivity and thermal stability, as reported in recent studies (Abid et al., 2015; Tiong et al., 2019). Fe_2O_3 is one of the most investigated nanoadditives within the cement matrix to produce corrosion-resistive cement because Fe_2O_3-modified cement showed less permeability even after corrosion, higher tensile strength, and less corrosion depth after the prolonged exposure to both CO_2 and H_2S (Bihua et al., 2018). The next class comprises nanobentonite, which can improve cement mechanical properties and reduce permeability compared to normal bentonite. Another class of nanoadditives is carbon nanomaterials providing high tensile strength and elasticity as well as thermal and electrical conductivity, for example, graphene nanoplates and carbon nanofibers (see El-Gamal et al., 2017). The next class is carboxylic acrylonitrile nanorubbers (Jafariesfad et al., 2017).

Polymeric additives can enhance cement resistance in corrosive environments. An important class of epoxy resins can form a separate phase within the cement matrix, preventing CO_2-induced reactions. However, the use of resin has temperature and durability restrictions. Another class is some expandable polymers that are triggered by the temperature to swell and expand. Such polymers are designed based on well conditions to expand after placement but before curing to ensure filling all cavities. Another useful class of polymer additives is biopolymers such as amine groups. Polymers also provide cement with a resilient and self-healing characteristic. As an example, superabsorbent polymers make cement swell to fill the microcracks inside the cement matrix, preventing any possible fluid channeling. As they absorb some water during the hydration process, affecting the water/cement ratio, a modified type should be essentially developed.

12.5 Geopolymers and improved integrity

Geopolymers are inorganic polymers composed of silicon (Group 4 of the periodic table) and aluminum (Group 3) as their main elements, which are activated with alkali solutions during the polymerization process, yielding polymers with Si–O–Al–O bonds, which can act as cement with less carbon footprint and thereby show great potential for becoming the replacement material for cementing in special wells in hostile environments. Geopolymer is not a very new topic: it goes back to 1972, when, in the aftermath of various catastrophic fires in France, causing hundreds of casualties in public buildings that involved common organic plastic, research on nonflammable and noncombustible plastic materials became a priority, and geopolymers were developed. However, the term "geopolymer" was first used by Joseph Davidovits in 1978. The chemical composition of geopolymers is similar to that of zeolites, but they exhibit an amorphous microstructure. The reaction leading to GPC creation is given in the following equation (Davidovits, 1991):

$$CaCO_3 + SiO_2 \rightarrow CaO \bullet SiO_2 + CO_2 \tag{12.2}$$

Volcanic ash and calcined clay/metakaolin are the two most common raw or base materials in geopolymers, which can become activated using strong alkaline solutions/activation models in a chemical process called polymerization. However, other base materials may be used, either with different geological origins or wastes. The use of clays has the requirement of energy consumption in a calcination process up to ~750°C; however, it is still more environmentally friendly than OPCs as the consumption rate is lower than that of OPC because of its significantly lower temperatures as compared to OPC clinker production (~1450°C). Fly ash is a waste material to be used for geopolymer production. However, as it is a by-product of burning coal for energy production, and coal burning is not actually environmentally friendly, this work does not necessarily promote using fly ash for GPC production, but it allows the opportunity of using this waste material if it is available.

Geopolymers for concrete use potentially present exceptional properties: being environmentally friendly (Davidovids, 2015), fire and heat resistant, acid resistant, having high early compressive strength, lower shrinkage, a good fracture toughness, and long-term durability. In addition to such good mechanical and physical properties, geopolymers seem much more sustainable than OPCs, owing to less energy required and less CO_2 emissions. Note that there is some minimal carbon footprint associated with geopolymer material production because of using the alkali activators because, for example, the embodied energies of sodium hydroxide (NaOH) and sodium silicate (Na_2SiO_3) are 14.9 and 5.4 GJ/t, respectively. However, as the required content for activators used is usually low (generally <10% of the total mix design), the carbon footprint with GPCs is considerably lower than OPCs; thus environmental benefits of GPCs are far greater than OPCs.

As for some works in industries other than construction, there are few publications in the petroleum industry, mainly focused on investigating the feasibility of using the GPC technology for well cementing (da Silva et al., 2023; Mario et al., 2023; Ogienagbon et al., 2023; Sazali et al., 2023; Zulkarnain et al., 2023). These works show that GPCs have comparable or sometimes better performance than OPCs in terms of strength and sealability. An example of the advantage would be OPC's permeability range is 0.001–0.15 mD, whereas GPC's permeability can be much lower: $4 \times 10^{-6} - 4 \times 10^{-5}$ mD (Nasvi et al., 2014). Therefore GPCs are ideal for plug and abandonment where the required material permeability for caprock hydraulic isolation is $10^{-6} - 10^{-3}$ mD (Khalifeh & Saasen, 2020). GPCs are also ideal for high-pressure and high-temperature (HPHT) wells (Salehi et al., 2017; van Oort et al., 2019). Abdelaal et al. (2023) confirmed the feasibility of using GPCs for making heavy cements of 17.5 ppg at 3000 psi and 292°F (145°C). The interesting point was that the slurry formulation showed a very low filtration loss.

Despite the advantages of geopolymers, such works are recent and are predominantly only in the research phase and on a small scale. There are multiple uncertainties in all these works as for comparison of OPC and GPC in terms of sealability and mechanical properties appropriate for well cementing, which include "shear bond strength," "hydraulic bond strength," volume shrinkage during setting, ballooning effects, casing diameter, pressure effect, temperature effect, and the CO_2 effect on rheological and mechanical properties of GPC's slurry and sheath/hardened samples. As an example, Gomado et al. (2022) found shear bond strength with GPC was extremely lower than that of OPC; however, they debated that because GPC showed to have almost identical hydraulic bond, the high shear bond may not be a full representation of complete sealability, and they therefore asked for more research. There are some works, such as Alanqari et al. (2022), for combining geopolymer and resin to make cements with improved properties.

No published works, to the best of authors' knowledge, have been found on the effect of Poisson's ratio as a crucial mechanical property indicating resilience of the cement, whereas this parameter should be investigated via detailed experimental mechanical tests. Mechanical tests are essential to be done at the interfaces between the cement and casing and formation for different types of OPCs and GPCs, and comparisons should be performed for feasibility and clarification of the right GPC's properties to be useful at HPHT and also in corrosive environments.

12.6 CO_2 evaluation techniques

Several evaluation techniques are available to assess cement integrity when exposed to CO_2. This section provides a brief overview of them.

12.6.1 Fiber optic sensors

Fiber optic sensors allow the detection of changes in acoustic waves along the wellbore. It has proven highly effective in identifying microfractures, leaks, and other anomalies in real time. In the context of CCUS, this technique is particularly valuable for monitoring well integrity, detecting dynamic strain along casing, identifying poorly bonded cement intervals, and observing seismic activity during and after CO_2 injection. Its main strengths include the ability to deliver high-resolution, continuous data in real time without requiring active sources, as well as being more cost-effective. However, it has some limitations; for instance, its radial resolution is limited, which can hinder the accurate localization of events occurring further from the cable. In addition, fiber degradation may occur under extreme downhole conditions (Raab et al., 2019).

12.6.2 Microcomputed tomography and nanocomputed tomography

Advanced imaging techniques are employed to identify and quantify changes in mineralogy and porosity resulting from the interaction of CO_2 with wellbore materials. In the context of CCUS, these technologies are applied to detect early signs of cement degradation or mineral precipitation, both of which can compromise the long-term integrity of the seal. By providing high-resolution 3D voxel-based imaging, microcomputed tomography and nanocomputed tomography enable nondestructive evaluation of core samples under conditions that simulate the in situ reservoir environment. Furthermore, they support the creation of digital twins of cement and rock samples, allowing simulations of mechanical strength, permeability evolution, and CO_2 flow dynamics (Chavez Panduro et al., 2021; Jahanbakhsh et al., 2021; Udebhulu et al., 2023).

12.6.3 Permeability testing while exposing the cement sheath to aqueous CO_2

Measuring permeability is essential for evaluating both well deliverability in production wells and injectivity in CO_2 storage applications. It plays a critical role in assessing the integrity and effectiveness of cement sheaths, helping to detect potential leakage pathways and weak zones within the wellbore structure (Metz et al., 2005; Udebhulu et al., 2023).

12.6.4 Mechanical testing while exposing the cement sheath to aqueous CO_2

Compressive strength and elastic modulus tests help assess the structural stability of cement after chemical interaction with CO_2. These tests reveal potential weakening of the cement matrix, development of microcracks, and changes in stiffness, all of which may compromise the sealing capacity and durability of the cement barrier over time (Udebhulu et al., 2023).

References

Abdelaal, A., Elkatatny, S., & Abdel Fattah, A. (2023). A geopolymer system for high pressure oil and gas well cementing. In *Paper presented at the Middle East Oil, Gas and Geosciences Show, Manama, Bahrain* February 2023 https://doi.org/10.2118/213661-MS.

Abid, K., Gholami, R., Choate, P., & Nagaratnam, B. H. (2015). A review on cement degradation under CO_2-rich environment of sequestration projects. *Journal of Natural Gas Science and Engineering*, *27*, 1149–1157. https://doi.org/10.1016/j.jngse.2015.09.061.

Ahmed, S., Ezeakacha, C. P., & Salehi, S. (2018). Improvement in cement sealing properties and integrity using conductive carbon nano materials: From strength to thickening time. In *SPE Annual Technical Conference and Exhibition*. Society of Petroleum Engineers https://doi.org/10.2118/191709-MS.

Alanqari, K., Al-Yami, A., & Wagle, V. (2022). Synthesis of epoxy resin based geopolymer utilizing Saudi Arabian Volcanic Ash for primary well cementing applications. In *Paper presented at the ADIPEC, Abu Dhabi, UAE* October 2022 https://doi.org/10.2118/211765-MS.

Bihua, X., Bin, Y., & Yongqing, W. (2018). Anti-corrosion cement for sour gas (H_2S-CO_2) storage and production of HTHP deep wells. *Applied Geochemistry, 96*, 155–163. https://doi.org/10.1016/j.apgeochem.2018.07.004.

Carey, J. W., Wigand, M., Chipera, S. J., WoldeGabriel, G., Pawar, R., Lichtner, P. C., Wehner, S. C., Raines, M. A., & Guthrie, G. D. (2007). Analysis and performance of oil well cement with 30 years of CO_2 exposure from the SACROC Unit, West Texas, USA. *International Journal of Greenhouse Gas Control, 1*, 75–85. https://doi.org/10.1016/S1750-5836(06)00004-1.

Chavez Panduro, E.A., Cordonnier, B., Gawel, K., Børve, I., Iyer, J., Carroll, S.A., Michels, L., Rogowska, M., McBeck, J.A., Sørensen, H.O., Walsh, S.D.C., Renard, F., Gibaud, A., Torsæter, M., & Breiby, D.W. (2020). Real-Time 3D Observations of Portland Cement Carbonation at CO_2 Storage Conditions. Environmental Science & Technology, 54(13), 8323–8332. doi:10.1021/acs.est.0c00578.

Da Silva, R. R., de Oliveira Freitas, J. C., Moreira, R. P. M., Braga, R. M., & Khalifeh, M. (2023). Development of a rock-based geopolymer for well abandonment applications—Utilizing Brazilian Rock Precursors. In *Paper presented at the Offshore Technology Conference Brasil, Rio de Janeiro, Brazil* https://doi.org/10.4043/32919-MS.

Davidovits, J. (1991). Geopolymers. *Journal of Thermal Analysis, 37*, 1633–1656. https://doi.org/10.1007/BF01912193.

Davidovits, J. (2015). False Values on CO_2 Emission for Geopolymer Cement/Concrete, Published in Scientific Papers, Technical paper 24, 1–9.

El-Gamal, S. M. A., Hashem, F. S., & Amin, M. S. (2017). Influence of carbon nanotubes, nanosilica and nanometakaolin on some morphological-mechanical properties of oil well cement pastes subjected to elevated water curing temperature and regular room air curing temperature. *Construction & Building Materials, 146*, 531–546. https://doi.org/10.1016/j.conbuildmat.2017.04.124.

Geraci, M., Jehangeer Ali, S., Romolt, C., Rossmann, R., & Rossmann, R. (2017). *The environmental risks and oversight of enhanced oil recovery in the United States: An overview of Class II well trends and regulations in EPA's Underground Injection Control Program*. Clean Water Action.

Gomado, F. D., Khalifeh, M., Kamali, M., Saasen, A., & Aasen, J. A. (2022). Sealing performance of geopolymer for plugging and abandonment; apple-to-apple scenario. In *Paper presented at the SPE Norway Subsurface Conference, Bergen, Norway* https://doi.org/10.2118/209552-MS.

Isahak, W. N. R. W., Ramli, Z. A. C., Ismail, M. W., Ismail, K., Yusop, R. M., Hisham, M. W. M., & Yarmo, M. A. (2013). Adsorption–desorption of CO_2 on different type of copper oxides surfaces: Physical and chemical attractions studies. *Journal of CO_2 Utilization, 2*, 8–15. https://doi.org/10.1016/j.jcou.2013.06.002.

Jahanbakhsh, A., Liu, Q., Hadi Mosleh, M., Agrawal, H., Mubeen Farooqui, N., Buckman, J., Recasens, M., Maroto-Valer, M., Korre, A., & Durucan, S. (2021). An Investigation into CO_2–Brine–Cement–Reservoir Rock Interactions for Wellbore Integrity in CO_2 Geological Storage. Energies, 14(16), 5033. DOI:10.3390/en14165033.

Jafariesfad, N., Khalifeh, M., Skalle, P., & Geiker, M. R. (2017). Nanorubber-modified cement system for oil and gas well cementing application. *Journal of Natural Gas Science and Engineering, 47*, 91–100. https://doi.org/10.1016/j.jngse.2017.10.002.

Jafariesfad, N., Sangesland, S., Gawel, K., & Torsæter, M. (2020). New materials and technologies for life-lasting cement sheath: A review of recent advances. *SPE Drilling & Completion*. https://doi.org/10.2118/199885-PA.

Khalifeh, M. & Saasen, A. (2020). Introduction to Permanent Plug and Abandonment of Wells, Springer International Publishing, https://doi.org/10.1007/978-3-030-39970-2.

Kiran, R., Teodoriu, C., Dadmohammadi, Y., Nygaard, R., Wood, D., Mokhtari, M., & Salehi, S. (2017). Identification and evaluation of well integrity and causes of failure of well integrity barriers (a review. *Journal of Natural Gas Science and Engineering*, *45*, 511–526. https://doi.org/10.1016/j.jngse.2017.05.009.

Mario, A., Salem, S., Tantawy, M., & Rammah, S. (2023). Utilizing innovative nanolime-geopolymer as a cement for gas wells. In *Paper presented at the ADIPEC, Abu Dhabi, UAE* October 2023 https://doi.org/10.2118/217002-MS.

Metz, B., Davidson, O., de Coninck, H., Loos, M., & Meyer, L. (Eds.). (2005). IPCC Special Report on Carbon Dioxide Capture and Storage (SRCCS). Cambridge, UK/New York, NY: Cambridge University Press.

Nasvi, M. C. M., Ranjith, P. G., Sanjayan, J., & Bui, H. (2014). Effect of temperature on permeability of geopolymer: A primary well sealant for carbon capture and storage wells. *Fuel*, *117*(Part A), 354–363. https://doi.org/10.1016/j.fuel.2013.09.007.

Nelson, E., & Guillot, D. (2006). *Well cementing* (2nd ed.). Schlumberger.

Ogienagbon, A., Korsnes, R. I., Khalifeh, M., Wu, Y., & Salehi, S. (2023). Performance of geopolymers in geothermal wells. In *Paper presented at the 57th U.S. Rock Mechanics/Geomechanics Symposium, Atlanta, Georgia, USA* https://doi.org/10.56952/ARMA-2023-0701.

Raab, T., Reinsch, T., Aldaz Cifuentes, S.R., & Henninges, J. (2019). Real-Time Well-Integrity Monitoring Using Fiber-Optic Distributed Acoustic Sensing. SPE Journal, 24(05), 1997–2009. https://doi.org/10.2118/195678-PA">https://doi.org/10.2118/195678-PA.

Salehi, S., Khattak, M. J., Bwala, A. H., & Karbalaei, S. ,F. (2017). Characterization, morphology and shear bond strength analysis of geopolymers: Implications for oil and gas well cementing applications. *Journal of Natural Gas Science and Engineering*, *38*, 323–332. https://doi.org/10.1016/j.jngse.2016.12.042.

Sazali, Y. A., Riyanto, L., Ebining, M. S., Rahman, S. H., Zulkarnain, N. N., Kumar, A. K., & Lau, C. H. (2023). Geopolymer for oilfield application: Scaling up laboratory test to yard test. In *Paper presented at the SPE Annual Technical Conference and Exhibition, San Antonio, Texas, USA* October 2023 https://doi.org/10.2118/215163-MS.

Tiong, M., Gholami, R., & Rahman, M. E. (2019). Cement degradation in CO_2 storage sites: A review on potential applications of nanomaterials. *Journal of Petroleum Exploration and Production Technology*, *9*, 329–340. https://doi.org/10.1007/s13202-018-0490-z.

Udebhulu, O. D., Aladeitan, Y., Azevedo, R. C., & De Tomi, G. (2023). A review of cement sheath integrity evaluation techniques for carbon dioxide storage. *Journal of Petroleum Exploration and Production Technology*. https://doi.org/10.1007/s13202-023-01697-0.

Van Oort, E., Juenger, M., Liu, X., & McDonald, M. (2019). Silicate-activated geopolymer alternatives to Portland cement for thermal well integrity. In *Paper presented at the SPE Thermal Well Integrity and Design Symposium, Banff, Alberta, Canada* November 2019 https://doi.org/10.2118/199787-MS.

Zhang, M., & Bachu, S. (2011). Review of integrity of existing wells in relation to CO_2 geological storage: What do we know? *International Journal of Greenhouse Gas Control*. https://doi.org/10.1016/j.ijggc.2010.11.006.

Zulkarnain, N. N., Hamid, Ab, Izwan, A., Sazali, Y. A., Razak, A., Amirhilmi, A., & Habarudin, M. F. (2023). The invention of a predictor plot for evaluating the response of the compressive strength of geopolymer cement to changes in temperature and pressure. In *Paper presented at the SPE/IATMI Asia Pacific Oil & Gas Conference and Exhibition, Jakarta, Indonesia* October 2023 https://doi.org/10.2118/215467-MS.

CHAPTER

13

High-pressure high-temperature and geothermal cement integrity evaluation

Jocin J. Abraham[a], Mahmood Amani[a], and Rahman Ashena[b]

[a]Texas A&M University at Qatar, Doha, Qatar, [b]Smart Energy Data Analytics (SEDA)-Group, Houston, TX, United States

13.1 Introduction

Due to running out of hydrocarbons in shallower oil and gas reservoirs and in pursuit of deeper reserves, there is an increasing trend for High Pressure High Temperature (HPHT) and deepwater wells. HPHT wells are defined as wells with temperature greater than 300°F (~150°C) and pore pressure of formation either greater than 10,000 psi or its maximum anticipated pore pressure gradient of the formation exceeding 0.8 psi/ft necessitating surface equipment pressure rating exceeding 10,000 psi. Drilling in HPHT zones has traditionally been considered challenging due to the complexities associated with hostile conditions including high pressure, temperature, narrow safe mud window, corrosive formation substances, etc. Hinton (1999) mentioned that HPHT wells were considered more risky than non-HPHT wells and constituted 22% of wells in offshore UK with annular flows during well construction. In fact, in the past and even now in some regions, HPHT zones may be avoided given the costly personnel and equipment requirements for planning, equipment, and execution. However, with further technological advancement, it is possible to drill and produce from such wells. Fig. 13.1 shows the Central Graben in North Sea where most HPHT wells are situated.

Well Cement Integrity
DOI: https://doi.org/10.1016/B978-0-443-51564-4.00006-0

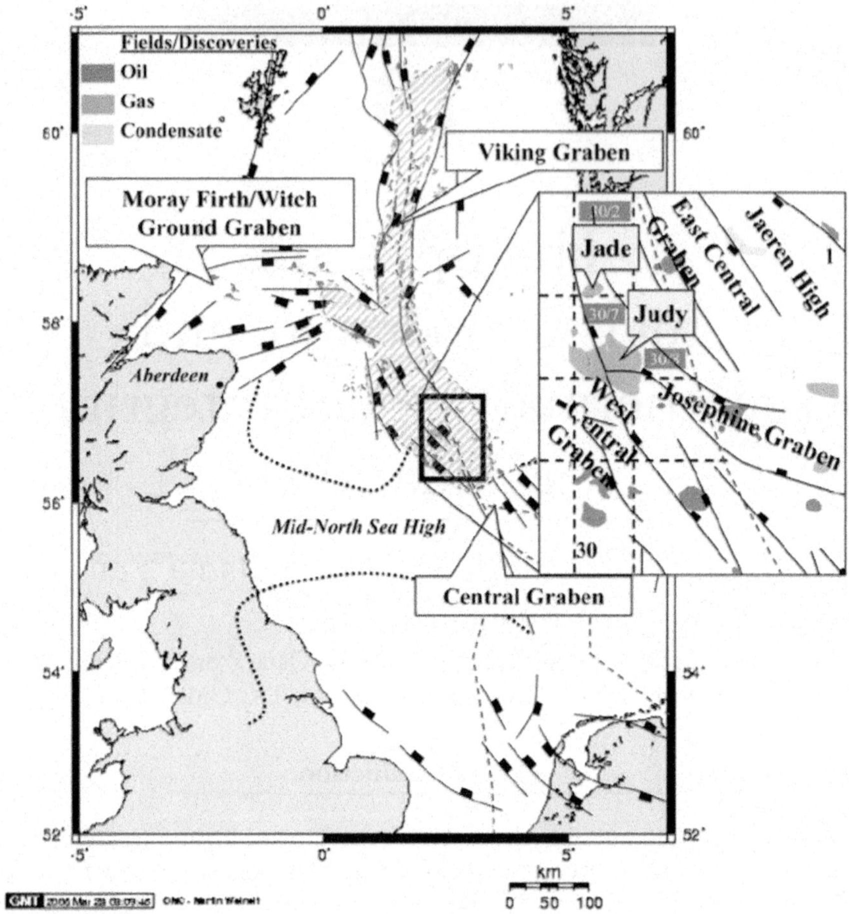

FIGURE 13.1 UK Central Graben gas reserves necessitating high-pressure high-temperature wells (Di Primio & Neumann, 2008).

The next hot topic nowadays is geothermal (GT) wells aiming at diversification of the energy provision in the energy basket beside that from oil and gas. A method of reaching this target is GT energy exploitation, necessitating drilling and producing wells in such reservoirs. The GT reservoir/source types may be classified as conventional and unconventional. Conventional GT sources are located near active or dormant volcanic regions and are widely available throughout the earth, and the scale of these resources is huge. Unconventional GT is referred to as repurposing too-old, obsolete/nonactivated, or even already abandoned oil and gas wells for GT energy production. To make any GT energy project work for enhanced geothermal systems (EGS), there are three requirements:

✓ High-temperature GT reservoir zone with temperatures up to 750°F,
✓ Presence of fractures or good permeability in the reservoir zone, and
✓ Injection/Supply of water flowing through the GT zone.

GT wells usually face well integrity issues during drilling because of high temperatures, high-pressure zones, and highly fractured formations (Allahvirdizadeh, 2020). High temperature causes dehydration of cement in addition to inducing thermal stresses on the cement sheath. Unconventional GT does not usually fall into the high-temperature category (Ashena, 2022) unless the abandoned wells were already in HPHT conditions. There are two common methods of GT well exploitation: Closed-loop geothermal (CLG) systems and EGS each have their advantages and disadvantages, and there are many research and industry works currently ongoing to make this technologically feasible and economically viable (Ashena, 2023; Pilko et al., 2021).

As for the EGS system, it is considered more economically applied by the industry, and it is worth noting that cementing is typically not performed across the producing zones. EGS systems are usually open-hole because the EGS efficiency is based on maximizing reservoir contact to optimize or maximize heat energy production from the returning GT fluids (Ashena et al., 2024). In case cement is present across the reservoir zones in EGS, it can adversely affect circulation and production of the returning fluid because of reducing contact area and thus decreased porosity and permeability, causing choking of the production pathways. In any case, well integrity from the production casing upward is an undoubtable necessity to allow reliable GT production. However, for CLG systems, we might be able to state that cement integrity is of greater importance than the EGS systems. This is because in such systems the whole reservoir section should be properly cemented, and there is significant heat transfer from the reservoir to the wellbore, which results in cement failure, knowing the mechanism of which requires further research. As for retrofitting oil and gas wells for GT production, well cement integrity is the main challenge, which should be properly addressed. It should be noted that cement integrity is a lesser concern in GT wells than in carbon capture, utilization, and storage wells where excellent well integrity is required.

To ascertain well integrity for these high-temperature cases the performance of oil well cements at higher temperatures should be essentially acceptable. Therefore the design of cement systems for GT applications should not only account for the strength and stability of the cement in the short run but also throughout the operational well life. It is crucial to consider the interaction between the cement sheath and the formation rocks and fluids, bonding between the cement and the casing wall, and the corrosion and salinity impact on the cement and casing. It is highly

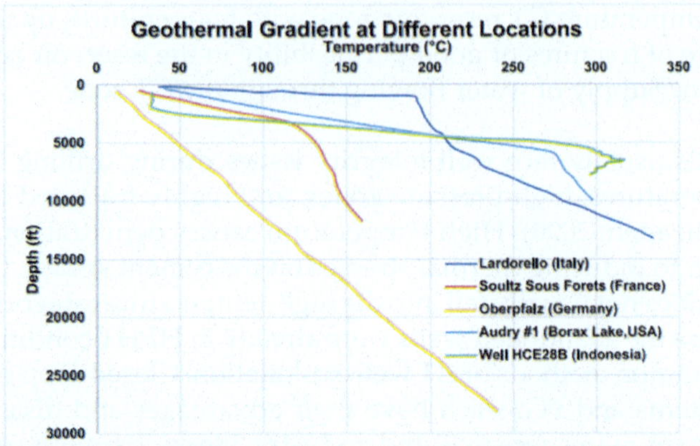

FIGURE 13.2 Geothermal gradients at different locations in the world, Adapted from Arbad et al. (2022).

important that cement slurries be reliably tested in the laboratory for a range of temperatures as present in HPHT/GT environments.

Generally, API-recommended classes of E, G, and H Portland cements may also be used for HPHT/GT applications. However, the cement slurry utilized may be modified by additives to account for and resist against the effects of downhole conditions, especially temperature and pressure, as well as formation fluids it will be exposed to (Fig. 13.2).

13.2 Affecting parameters and impact

13.2.1 Cyclic loading

Cement is considered a brittle material that can fail when subjected to repeated exertion and release of stress induced by well pressures, called cyclic loading. Therefore cement integrity is affected by cyclic loading, which occurs during well events such as hydraulic fracturing, squeeze cementing, and acidizing. Cyclic loading does not immediately cause damage, but it works in a course of time by which the cement material begins to accumulate damage, leading to subsequent formation of fissures and cracks and finally total cement failure. Therefore it is of great importance to evaluate a hardened cement's tolerance to cyclic loading in lab environment to ensure its integrity in situ. Fig. 13.3 shows CT scans of two cement samples in two different works after exposing them to cyclic loading: The first sample was Class H cement (with silica flour) sample, which failed under 12 cycles of 1000 psi after curing at 15,000 psi and 330°F loading, whereas the second sample was a Class G cement (samples i and iii

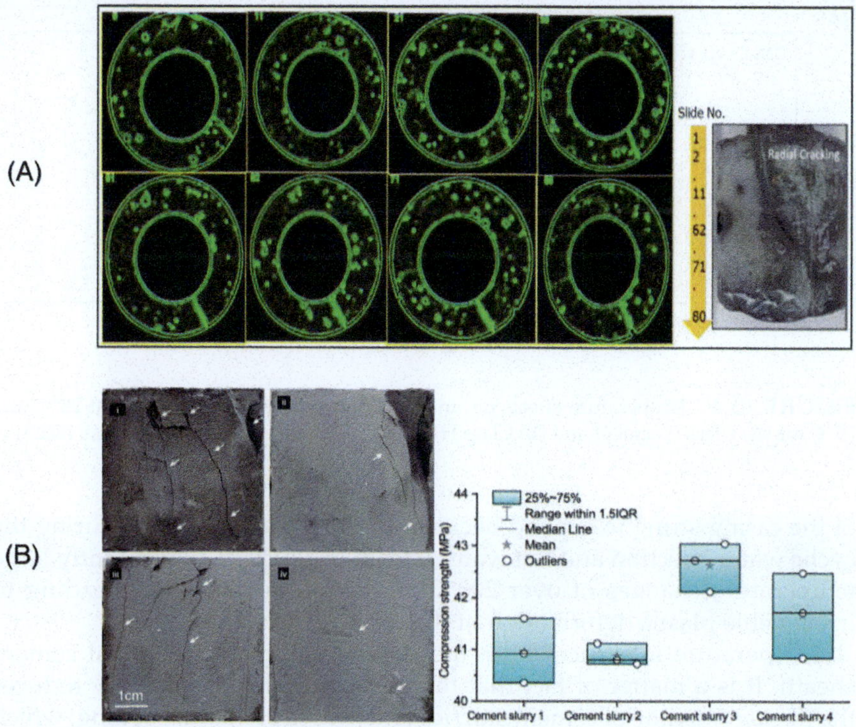

FIGURE 13.3 CT scans of: (A) a Class H cement sheath after exposure to high-pressure high-temperature cyclic loading showing cement failure (Shadravan et al., 2014), and (B) a cement reinforced by propylene fiber showing no cement failure, just a single radial crack that would not propagate (Hao et al., 2023).

without fiber, samples ii and iv with carbon and basalt fiber) reinforced by inorganic fibers (carbon, basalt), which remained almost integrated (with just a simple radial crack) under different load cycles after curing at 160°F.

13.2.2 Temperature

As a cemented casing in an HPHT/GT environment is generally subjected to temperature changes, cement chemical alteration as well as tensile and compressive stresses can develop in the cement sheath and casing. At temperatures above 230°F the calcium silicate hydrate (C–S–H) phase, which is responsible for the set cement's strength and dimensional stability, transforms to alpha dicalcium silicate hydrate (C_2SH). This phase is denser and more crystalline than the C–S–H phase. As a result, matrix shrinkage might occur, compromising the set-cement integrity (Dahi Taleghani & Santos, 2023). Stress development occurs mainly because of the inability

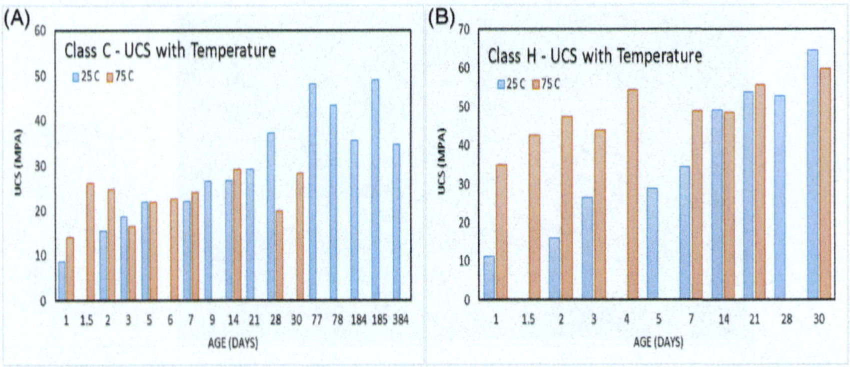

FIGURE 13.4 Temperature effects on uniaxial compressive strength (UCS) at 25°C and 75°C for (A) Class C cement and (B) Class H cement, Adapted from Abraham et al. (2023).

of the casing string to expand because of temperature changes during the cyclic water injection and hot water/steam production. Significantly large temperature changes of over 200°C can cause huge stresses, leading to irreversible plastic deformation of the cement sheath and casing.

Temperature influences both the cement slurry and the final cement sheath. It is a matter of fact that high temperature considerably reduces the thickening time of cement slurries and causes premature setting, which will in turn influence the cement sheath properties. Several cement sheath properties that are affected by temperature consist of uniaxial compressive strength (UCS), Young's modulus, Poisson's ratio, and tensile strength as well as bond strength. These would determine the cement ductility and its failure modes. It is a fact that cements should not be brittle; otherwise, under the hostile downhole conditions, they would fail through (1) microannulus failure mode, during which they debond from the casing or formation, leading to a microannulus generation, and/or (2) cracking failure, which means cracks may be created and in turn propagate to the cement boundaries, leading to entire deterioration of the cement.

The cement compressive strength is usually measured by the industry after 24 hours of curing in the curing chamber, with an expected range of 2500–2750 psi considered acceptable. An accelerated setting of the cement samples usually leads to the development of lower threshold compressive strengths; this was confirmed by Abraham & Amani's (2023) tests. It can be reasoned that at higher temperatures, cement is rapidly dehydrated thermally, causing its earlier setting while compromising on its peak compressive strength. Fig. 13.4 shows that API Class H and Class C cements cured in a thermally active environment had significantly lower thickening and hardening time. Peak compressive strengths were reached in 2–4 days for thermally cured cements, compared with those cured at room temperature, which typically took around 14–21 days. In addition, peak compressive strengths for thermally cured cement were 20%–40% lower than those cured at lower temperatures (Abraham et al., 2023). Note that Class C cement is mentioned here because many old wells, especially in the United States, were cemented using this cement class, and this type of cement would be encountered in old wells in the case of repurposing such wells to GT.

Many studies have already mentioned cement strength degradation over time because of exposure to high temperatures; see Arbad et al. (2022). MacAndrew et al. (1993) showed that the compressive strength of the set cement deteriorates with time, and it is more likely once the expected bottomhole static temperature (BHST) exceeds 225°F (~107°C). Studies by Carter and Smith (1958) and Saunders and Walker (1954) have shown that above critical temperatures of 230°F (110°C), the cement strength showed a decrease over time (for tests on samples aged 28–180 days) with an increase in temperature for certain formulations, a phenomenon referenced to as cement strength retrogression. This phenomenon was more pronounced at higher temperatures and depends on the aging time and cement slurry composition.

In addition to the formation temperature the set cement may encounter another source of temperature. During production operations, high-temperature steam may be injected to flow through the production casing string. Several studies in HPHT/GT production systems have documented that hardened cements undergo a phenomenon referred to as "strength regression" when exposed to high temperatures (>105°C/221°F). This signifies that the already set and hardened cement will lose its UCS because of exposure to high temperatures over time, leading to cracking and cement failure situations (Rachimillah et al., 2021). In GT wells the cycle of cold (low temperature) and hot (high temperature) water injection causes differential stresses on the cement, causing degradation of the cement and detaching the cement bond from the casing causing the formation of microannulus (Therond et al., 2016). The cyclic temperature changes because of pumping and production operations common in enhanced GT Systems can exacerbate this issue, accelerating such failure in the cement systems.

For successful HPHT/GT cementing, proper cement and cementing designs should be made to have cements that can withstand significantly high stresses during the operational well life and follow best cementing practices that encourage utilizing excess cement when completing GT wells (Arbad et al., 2022). For a good design the first step is to estimate temperatures in both static mode BHST and dynamic mode (bottomhole circulating temperature, BHCT) prior to execution of cementing for cement and cementing design. This estimation is doable either during drilling by measurement while drilling or logging while drilling sensors or sensors used in the cleanup trips, or if none are available, using computer-based thermal modeling programs. There are some differences between all these measurements. The BHCT is useful for cement slurry design, and BHST is useful for cement behavior prediction/modeling for the setting and hardening phase.

To resolve this issue of cement strength degradation a common point to consider in cement design is to use silica flour as a stabilizing agent to counteract the effect of decline in the UCSs of cement downhole. Several studies have mentioned amounts of silica to be added as an additive for different temperatures downhole, and among the recommendations, Arbad et al. (2022) used an optimal amount of 15%–20% silica flour, by weight of cement (bwoc). In the industry, 32–35 lb/sack is usually used for deep oil and gas drilling. This would need to be further optimized for different temperature gradients typically experienced downhole in GT wells and account for the temperature changes during injection and production operations when water and superheated steam are cycled through the well. Usually, in such operations with temperatures exceeding 230°F, extra silica up to 40% (bwoc) may be recommended.

As mentioned in the previous chapter, geopolymer cements have ideal properties, including high-temperature resistance, high strength, and sealability compatible with high temperature cases.

13.2.3 Cement bond

In designing cements for GT applications, along with the strength and stability criteria, another critical factor to be considered is the cement bond to ensure wellbore isolation and maintain production integrity. The cyclic injection and production of superheated fluids into and from the formation cause rapid thermal expansion and compression of the casing and cement layer, and these stresses can in turn cause the cement to crack and debond at the casing-cement or the cement-formation interfaces.

Ductility/Resilience of cement delineates the resilience of the cement in terms of its reaction to high variations of temperature and pressures in the wellbore and its ability to adapt and create sealing again. To be more exact the thermal expansion coefficient of the casing is less than 20% that of the cement, causing incomparable expansion of cement and casing, causing incomparable expansion of casing versus cement. Thus, if the cement is brittle, it cannot be flexible enough in facing high variation of pressures and temperatures and would debond from the casing, leading to the creation of either microannulus or, even worse, microchannels. Therefore a cement sheath should have an acceptable Poisson's ratio of say 0.15–0.25. Poisson's ratio is not usually conducted by the oil and gas industry on their cement samples; thus more research is required on this topic. Ahmed (2018) found that concrete Poisson's ratio decreased sharply for the first 24 hours from 0.5 (fluid-like matter) to 0.08 and then rebuilt to 0.15 in 7 days.

Several studies have shown the impact of rock types on cement interface bonding. Cerasi et.al. (2020) showed that typical Portland cement formed a better bond with a sandstone rock compared with the bonding with a shaly rock. Cement bond strength is also impacted by the presence of a mud cake on the formation. The thickness and coverage of the mud cake determined the bonding efficiency and the bond strength that developed between the cement and the formation. Likewise the presence of a thin mud layer/film on the casing-cement interface would play a negative role in determining the quality of the cement bond with the casing surface. A 10% mud contamination with cement may cause about a 23% reduction of cement strength after 16 hours (MacAndrew et al., 1993).

To ensure proper bonding, especially in the high-temperature zones, it is recommended to include cement additives or expanding agents like MgO to improve the cement bonding at the interfaces (Buntoro and Rubiandini, 2000). It is also recommended that routine bonding and rebonding tests be conducted to ensure adequate cement bonds and test the self-healing properties of the cement. If inadequate, remedial or squeeze cementing should essentially be carried out to provide cement in uncemented intervals to restore cement integrity to proper condition.

13.2.4 Corrosion

Acidic corrosive gases, such as CO_2 and H_2S, are also very commonly present in HPHT/GT wells owing to their proximity to high-temperature zones. As for the casing, at high temperatures (>185°F/85°C), corrosive gases do not affect steel; however, at lower temperatures, they attack steel. The solution favored by many North Sea operators is to use thick-walled, special-grade steel 9 7/8-in or even 10 3/4-in casing higher up the well tapering to 9 5/8-in casing.

As for the cement, as the base of Portland cement is limestone and silica, it is prone to attacks from CO_2 and H_2S, which can cause loss of the structure and stability of the cement sheath and thus the loss of well integrity, especially in the long term.

Several studies have investigated the effect of CO_2 on cement rock interactions and methods to mitigate it. Acid attacks caused because of the interaction between cement and CO_2 through a carbonation reaction degrades the cement, transforming it to calcite and aragonite (Arbad et al., 2022). Casing leaks can often develop because of this, especially in the connections or in zones where the cement job is inadequate or degraded. A very common form of this is a leak in the tie-back receptacle in a casing string, and it typically occurs when there is inadequate cement coverage between the tieback and liner. To rectify this issue, it is often recommended to increase the cement coverage by pumping additional high-viscosity cement into the tie-back liner overlap (Rachimillah et al., 2021).

It is a common practice to add silica flour/powder (sodium silicate and Na_2SO_3) as a stabilizing agent to increase the cement's compressive strength and decrease its permeability. This, in turn, can make the problem of corrosion in cement more manageable as the rate of carbonation is significantly slowed owing to the lower cement permeability. Industry and research laboratories have also developed alternatives to conventional Portland cement, which can be lightweight and can also resist acid attacks because of CO_2 and H_2S. Some common alternatives already tested to resist acid attacks at temperatures up to 320°C are the "ThermaLock Cement" from Halliburton, "calcium aluminate phosphate (CAP)," and "sodium silicate–activated slag (SSAS)" cement recipes from the Brookhaven National Laboratory (Arbad et al., 2022).

13.2.5 Salinity

Cement slurries used in GT applications must be tested for compatibility with formation water, both with and without free water present. At higher temperatures near the GT zone, free formation water is typically absent, but closer to the surface or in zones above the GT, free formation water may be present between the casing, cement, and formation layers. In addition, formation water in GT zones may have higher salinity than in other parts of the well. Therefore cement slurry testing should account for all wellbore conditions, including fluids and temperatures along the entire length of the well. It is also recommended that laboratory tests use the same formation water for aging to better simulate actual GT well conditions.

13.3 Other cementing related integrity challenges

13.3.1 Isolation issues

One of the main challenges in GT wells is isolating a high-temperature zone because of the fact that during cementing operations typical permanent or retrievable bridge plugs cannot be used. This lack is because of the operational parameters exceeding their optimal temperature ratings as well as potential degradation of the rubber sealing elements. It is a matter of fact that a damaged plug can cause isolation issues and difficulties in milling through the isolation zone as well as potential plugging of the steam flow paths once it drops to the producing zones. To avoid the mentioned issues, it is typically recommended to utilize wooden plugs for isolation in such zones. The wooden plug is typically dropped on top of the liner, and then cement is placed on top of it as a temporary barrier to isolate the zone. Once the workover operation is completed, the cement and wooden plug can be milled out before resuming GT operations (Rachimillah et al., 2021).

13.3.2 Casing deformations—Cold collapse

In HPHT/GT wells, inadequate or faulty cementing can result in water being trapped in the annulus between casing-casing or in isolated channels or pockets between the cement and casing layers. The trapped fluid pockets can expand considerably as the temperature increases during the production of superheated steam and fluids from such wells, leading to an annular pressure buildup or sustained annular pressure. This can in turn exert a force on the casing, causing damage to the walls of the casing and causing it to collapse when the force exceeds the collapse rating of the casing string (Rachimillah et al., 2021).

Commonly referred to as "cold collapse," this failure occurs commonly in GT wells and is a combination of high tension and excessive collapse pressures acting on the tubulars during repeated heat cycles of pumping superheated fluids and cooling down during shut-in, called quenching (Suryanarayana et al., 2021). As per an earlier approximation by Torres (2014), every 1°F rise in temperature will result in an increase in pressure of about 50 psi for trapped water/steam. This problem can be mitigated to an extent by utilizing higher grade casing and tubing strings with higher collapse pressure ratings, but this is not sustainable, as higher grade casing strings are more expensive, and with the production of superheated fluids, the pressures can eventually exceed even the highest rated collapse pressures.

Another common issue which is worsened in GT wells due to patchy or inadequate cement placement is the increase in the buckling tendency and casing deformation. If there is a significant length of the uncemented/faulty cemented section, the buckling can be inelastic due to the non-uniformity of the forces acting on the casing string. Inelastic buckling reduces the maximum allowable temperature of the casing string and can also cause failures in tubular connections. Depending on the direction of the differential pressure acting on the string, this buckling can be inwards, which can lead to a local collapse of the string, or bulge outward, which can lead to a burst pipe as shown in Fig. 13.5. It is therefore very critical that all major sections of the casing string are adequately cemented, or a higher grade of pipe be utilized to account for these issues (Suryanarayana et al., 2021).

13.3.3 Cold water intrusion

Poor cementing and failures in well isolation can result in a loss of thermal capacity of the superheated steam, leading to reduced energy production efficiency in GT systems. This typically occurs when water seeps into the borehole from near the surface or collects behind the casing string. As these leaks happen at shallower depths closer to the water

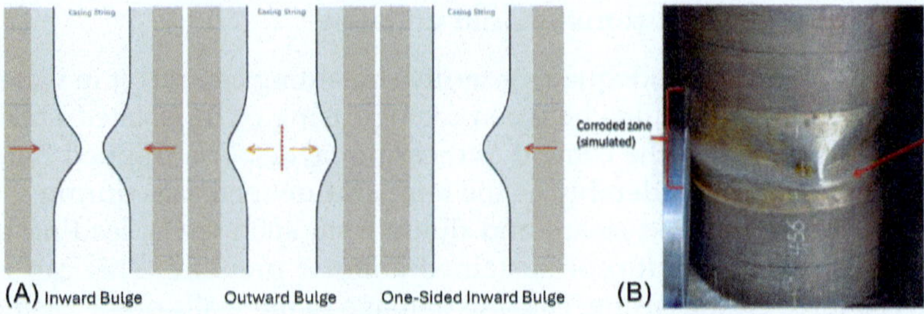

FIGURE 13.5 (A) Schematic of inward or outward buckling, Adapted from Suryanarayana et al. (2021) and (B) Deformation of a casing string due to buckling (Teodoriu, 2015).

table, where temperatures are lower (cold), the resulting interactions can decrease the steam temperature and reduce energy production efficiency.

13.3.4 Lost circulation

GT wells, particularly the EGS ones, which are usually hydraulically fractured, are subject to massive lost circulation issues during drilling because of the presence of natural or induced fractures in the formations, the presence of vugs or cavernous formations, and the general volcanic nature of the rocks. Remedial cementing is often advised for stabilizing problematic zones and curbing circulation losses before drilling ahead. During drilling into GT formations, lost circulation plugs are often needed to address circulation losses of both the drilling fluid and the cement slurry, or level-off plugs may be advised (Ashena et al., 2021).

For any circulation loss issue, it is necessary to understand the characteristics of the losses, as different losses need different solutions and cement slurry designs. Rachimillah et al. (2021) described some of the approaches for tackling, the issue of lost circulation based on its severity, and they are listed in Table 13.1. Special "reactive cements" can be utilized, which typically set very rapidly and can help decrease or stop the losses. A common example of this is silicate solutions such as silica flour and water, which form a strong, semipermanent gel when mixed with the cement slurry. This reaction causes the slurry to harden quickly, effectively sealing the lost circulation zone.

13.4 Cement bond log interpretation

One of the major challenges of using conventional cement bond log (CBL) tools in HPHT/GT wells is their limitation when it comes to the temperatures encountered in the well. Even the tools rated for HPHT/GT

TABLE 13.1 Loss of circulation cement approach.

Severity	Approach
Total Lost circulation condition (typically >500 bbl/h)	Pump higher density cement slurry, which has a fast gel strength development rate. Generally observed to increase the success rate of lost circulation cement plugs
Severe lost circulation conditions (typically 200–500 bbl/h)	A remedial cement job using reactive cement can mitigate or cure the circulation losses. Gaining fluid returns is required to continue drilling ahead, and reactive cements can be used for successful lost circulation cement plugs
Partial lost circulation conditions (typically 10–200 bbl/h)	Adjust the fluid density with respect to the expected fracture pressure to avoid increasing the rate of lost circulation

operations typically may not meet the threshold temperature requirements for such wells. A quick survey into some of the most common CBL tools employed by industry service providers SLB (2025) and Baker Hughes (2023) shows that the upper limit of safe operating temperatures for their tools is around 177°C (350°F), whereas temperatures in conventional GT wells may exceed 300°C (572°F).

Because of the mentioned temperature limitations of CBL tools, wells must often be cooled down before logging, and logging is suspended in those zones where temperatures exceed safe operating temperatures. However, when the temperature of the well is reduced, debonding can occur as the steel casing pipe contracts more than the cement sheath, thereby creating a "microannulus" effect between the cement and the pipe. Though this annulus effect is minimal, it can destroy the apparent bond registered by the CBL log and show up as no bond present between the casing and cement. As a result, most operators consider the validity of such logs obtained by running tools at lower temperatures to be doubtful (Chang, 1981). Fig. 13.6 is a CBL from the RN-15/DEEPEGS/IDDP-2 GT well in the Reykjanes GT field in southwest Iceland, which demonstrates the challenges of logging in GT zones. Cementing was performed on the well during drilling, and two back-to-back CBLs were run to monitor the evolution of the cement-casing bond over time as the cement sets. The tools used for CBL were temperature limited to 150°C, and so sections below 2500 m were not logged. Several zones along the well path showed temperature fluctuations, with higher temperatures seen at certain depths and cooling or lower temperatures zones at deeper depths, representing an inconsistent GT gradient. CBLs were run after 14–24 hours, and 33–40 hours and both showed inconsistencies and poor cement bond throughout most of the logged intervals (lower amplitude means better bond). The second CBL run showed a slightly better cement bond than the first, but

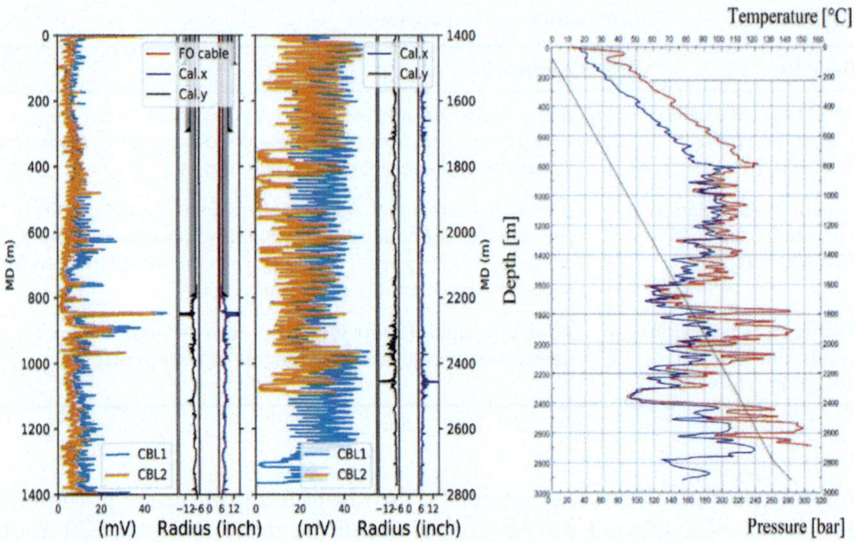

FIGURE 13.6 Cement bond logs (CBLs) for RN-15/DEEPEGS/IDDP-2 GT well in Iceland, CBL1 run after 14–24 hours. CBL2 run after 33–40 hours. Temperature and pressure for depth reference, Adapted from Lipus et al. (2021).

it was still nonuniform, with certain sections indicating a fully unbonded casing/free pipe. This is a typical challenge with high-temperature wells, as placement of cement can be nonuniform and the temperature variations can delay the setting and hardening of the cement (Lipus et al., 2021).

13.5 Recommended operational practices

- It is recommended that the API oil well cements used for GT operations be tested at various temperatures and for different periods. Therefore the setting and hardening properties of cements can be determined for different temperature ranges, which are likely to be present at different depths.
- To account for the cement strength regression because of the high temperatures encountered in GT environments, it is recommended to use additives like silica flour. Silica flour with concentration ranging from 15% to 20% can counteract most of the thermal degradation effect on the cement strength.
- In the case of highly corrosive environments, which may be present in HPHT/GT wells, again silica flour is already known to improve the cement resistance to CO_2 and H_2S. In addition, some industry-recommended alternatives may be used, such as CAP, SSAS, or "ThermaLock" cement, which are better suited to GT environments.

- In almost all cases, the policy of "more is good" should be essentially followed for cementing GT wells. Cementing throughout is recommended for wells drilled in high-temperature and corrosive GT environments. In other words, instead of cementing specific zones, as is the case with most operators to save cement cost especially in the United States, it is almost always better to cement the entire well when being used for GT or HPHT applications. Therefore a 100% coverage of the casing-formation or casing-casing annulus can ensure that formation leaks as well as temperature-related debonding issues are minimized.
- As HPHT/GT wells introduce temperature limitations on CBL tools, efforts must be made to make these evaluations at temperatures as close to reservoir and well temperature as possible.
- In case areas of debonding or microannuli presence are detected, remedial cementing or squeeze cementing must be essentially performed before starting production operations. High-temperature operations in HPHT/GT wells can exacerbate these issues and can cause casing and cement damage as well as efficiency and integrity issues.
- Apart from cement the well design should be strictly checked in terms of casing and tubular string selections for their burst and collapse resistance for GT applications. Note that safety factors for design considerations of tubulars must go above and beyond the recommended standards because of accelerated failure mechanisms by thermal conditions exceeding the maximum allowable range.

Abbreviations

HPHT	high-pressure high-temperature
CBL	cement bond log
CLG	closed-loop geothermal
CAP	calcium aluminate phosphate
DAS	distributed acoustic sensing
DTS	distributed temperature sensing
EGS	enhanced geothermal systems
UCS	uniaxial compressive strength
USIT	ultrasonic imaging tool
VDL	variable density log
SSAS	sodium silicate–activated slag

References

Abraham, J. J., & Amani, M. (2023). Field evaluation and assessment of wellbore, casing and cement integrity during critical high-pressure operations in existing and abandoned wells. In *Gas & oil technology showcase and conference*. Society for Petroleum Engineers https://doi.org/10.2118/214035-MS.

Abraham, J.J., Devers, C., Teodoriu, C., & Amani, M. (2023). A Comprehensive Evaluation of Properties of Common and Regional Cement Formulations from a Well Integrity Standpoint. Paper presented at the Offshore Technology Conference, Houston, Texas, USA, May 2023. https://doi.org/10.4043/32419-MS.

Ahmed, L. (2018). Dynamic measurements for determining Poisson's ratio of young concrete. *Nordic Concrete Research, 58*(1), 95–106. https://doi.org/10.2478/ncr-2018-0006.

Allahvirdizadeh, P. (2020). A review on geothermal wells: Well integrity issues. *Journal of Cleaner Production, 275*, 124009. https://doi.org/10.1016/j.jclepro.2020.124009.

Arbad, N., Emadi, H., & Watson, M. (2022). A comprehensive review of geothermal cementing from well integrity perspective. *Journal of Petroleum Science and Engineering, 217*. https://doi.org/10.1016/j.petrol.2022.110869.

Ashena, R. (2022). An Overview of Retrofitting Abandoned Petroleum Wells for Geothermal Energy Production. Paper presented at the ADIPEC, Abu Dhabi, UAE. https://doi.org/10.2118/210861-MS.

Ashena, R., Ghalambor, A., Elmgerbi, A. et al. (2021). 'Level-off' cement plugging method to cure lost circulation verified with case studies. *J Petrol Explor Prod Technol* 11, 2777–2789. https://doi.org/10.1007/s13202-021-01177-3.

Ashena, R. (2023). Analysis of some case studies and a recommended idea for geothermal energy production from retrofitted abandoned oil and gas wells. *Geothermics, 108*, 102634. https://doi.org/10.1016/j.geothermics.2022.102634.

Ashena, R., Ghalambor, A., Bataee, M., & Thiruvelcham, V. (2024). A workflow for circulation rate optimization to exploit enhanced geothermal system productivity. In *Paper presented at the SPE International Conference and Exhibition on Formation Damage Control* February 2024 https://doi.org/10.2118/217923-MS .

Baker Hughes (2023). *Acoustic cement bond log (CBL) service*. Baker Hughes https://www.bakerhughes.com/evaluation/wireline-well-integrity-evaluation/cement-evaluation/acoustic-cement-bond-log-cbl-service.

Buntoro, A. R. I. S, & Rubiandini, R. S. (2000). The effect of neat magnesium oxide as expanding additive on cement shear bond strength at high temperature up to 2500°C. In *World Geothermal Congress* http://www.geothermal-energy.org/pdf/IGAstandard/WGC/2000/R0425.PDF.

Carter, G., & Smith, D. K. (1958). Properties of cementing compositions at elevated temperatures and pressure. *Trans, 213*, 20–27. https://doi.org/10.2118/892-G.

Cerasi, P., Ghaderi, A., Taghipour, A., Todorovic, J., Gawel, K., Coquard, R., Edvardsen, L., Agofack, N., & Bhuiyan, H. (2020). Testing methods to assess improved well cementing and remediation. *Transactions – Geothermal Resources Council, 44*, 29–37.

Chang, H.T. (1981). Acoustic cement bond logging diagnostics for geothermal applications, article, January 1, 1981; Albuquerque, New Mexico. (https://digital.library.unt.edu/ark:/67531/metadc1208026/. Accessed March 7, 2026), University of North Texas Libraries, UNT Digital Library, https://digital.library.unt.edu; crediting UNT Libraries Government Documents Department.

Dahi Taleghani, A., & Santos, L. (2023). *Well integrity from theory to practice*. Springer https://doi.org/10.1007/978-3-031-19024-7.

Di Primio, R., & Neumann, V. (2008). HPHT reservoir evolution: A case study from Jade and Judy fields, Central Graben, UK North Sea. *International Journal of Earth Sciences, 97*, 1101–1114. https://doi.org/10.1007/s00531-007-0206-y.

Hao, H., Song, J., Chen, M., Yan, X., & Zhang, K. (2023). Rheological and mechanical properties of oil-well cement reinforced by hybrid inorganic fibers. *Construction and Building Materials, 377*, 131002. https://doi.org/10.1016/j.conbuildmat.2023.131002.

Hinton, A. (1999). An analysis of OSD's well incident database; results can improve well design and target well control training. In *SPE Offshore Europe oil and gas conference and exhibition* https://doi.org/10.2118/56921-MS.

Lipus, M. P., Reinsch, T., Weisenberger, T. B., Kragset, S., Stefansson, A., & Bogason, S. G. (2021). Monitoring of a reverse cement job in a high-temperature geothermal environment. *Geothermal Energy, 9*, 5. https://doi.org/10.1186/s40517-021-00187-y.

MacAndrew, R., Parry, N., Prieur, J., Wiggleman, J., Diggins, E., Guicheney, P., & Cameron, D. (1993). Drilling and Testing Hot High Pressure Wells. Oilfield Reviews 5 (2), 15–32.

Pilko, R. M., Hart-Wagoner, N. R., Horn, A. J. V., & Scherer, J. A. (2021). Repurposing oil & gas wells and drilling operations for geothermal energy. In *Offshore Technology Conference*. Society of Petroleum Engineers.

Rachimillah, A. P., Azwar, C., Johri, A., Osman, A., & Tanoto, E. (2021). Obtaining zonal isolation in geothermal wells. In *IADC/SPE Asia Pacific Drilling Technology Conference*. Society of Petroleum Engineers.

Saunders, C. D., & Walker, W. A. (1954). Strength of oil well cements and additives under high temperature well conditions. In *Fall Meeting of the Petroleum Branch, American Institute of Mining and Metallurgical Engineers* https://doi.org/10.2118/390-G.

Shadravan, A., Schubert, J., Amani, M., & Teodoriu, C. (2014). HPHT cement sheath integrity evaluation method for unconventional wells. In *Paper presented at the SPE International Conference on Health, Safety, and Environment* https://doi.org/10.2118/168321-MS.

SLB (2025). Cement Bond Logging Tools. SLB. https://www.slb.com/products-and-services/innovating-in-oil-and-gas/well-construction/well-cementing/cement-evaluation/cement-bond-logging-tools.

Suryanarayana, P. V., Kishnamurthy, R. M., Sathuvalli, U. B., & Bowling, J. (2021). A review of casing failures in geothermal wells. In *World Geothermal Congress 2020+1* https://www.geothermal-energy.org/pdf/WGC/papers/WGC/2020/21054.pdf.

Teodoriu, C. (2015). Why and When Does Casing Fail in Geothermal Wells: A Surprising Question? Proceedings, World Geothermal Congress 2015, Melbourne, Australia. https://worldgeothermal.org/pdf/IGAstandard/WGC/2015/21041.pdf.

Therond, E., Bois, A.-P., Whaley, K., & Murillo, R. (2016). Large scale testing & modelling for cement zonal isolation of water injection wells. In *Presented at the SPE Annual Technical Conference and Exhibition*. SPE Day 2 Tue, September 27, 2016 https://doi.org/10.2118/181428-MS D021S023R004.

Torres, A. (2014). Challenges of casing design in geothermal wells. In *IADC/SPE Asia Pacific Drilling Technology Conference*. Society of Petroleum Engineers IADC/SPE-170480-MS https://doi.org/10.2118/170480-MS .

Index

Page numbers followed by "*f*" and "*t*" indicate, figures and tables respectively.

M

N

O

P

R

S

Printed and bound by CPI Group (UK) Ltd, Croydon, CR0 4YY
11/09/2026
02215552-0001